LIVING IN SPACE

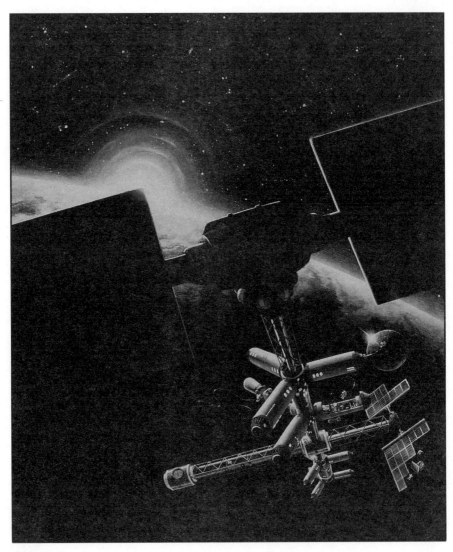

(NASA photo)

LIVING IN SPACE

A Handbook for Work & Exploration
Beyond the Earth's Atmosphere

G. HARRY STINE

M. EVANS AND COMPANY, INC.
New York

M. Evans and Company, Inc.
216 East 49th Street
New York, New York 10017

Library of Congress Cataloging-in-Publication Data

Stine, G. Harry (George Harry), 1928–
  Living in space / G. Harry Stine. — 1st ed.
    p. cm.
  Includes index.
  ISBN 0-87131-841-5
  1. Life support systems (Space environment) 2. Space colonization. I. Title.
TL1500.S75 1997
629.47'7—dc21   97-14880
                                                                       CIP

Design and composition by John Reinhardt Book Design

Manufactured in the United States of America

First Edition

9 8 7 6 5 4 3 2 1

To
Jennie and Laura
who may need this

CONTENTS

FIG. 1-1. *People can live and work in space, either in space facilities or using space suits. It can be done, it has been done, and it will continue to be done. The new frontier awaits more people.* (NASA photo)

CHAPTER ONE

SPACE IS FOR PEOPLE

FROM NOW UNTIL THE STARS GROW COLD, human beings will be living, working, playing, exploring, and traveling in space beyond the Earth's atmosphere.

We are the first species with the ability to leave Planet Earth and expand the horizons of existence into the infinite realm of the universe. Humanity may have been working, learning, and building toward this accomplishment throughout human history. Not everything people do or have done is a step toward this goal, but the long-term trends point toward what Robert A. Heinlein has called "the Great Diaspora."

This book will help people work, live, and play beyond the Earth's atmosphere. It covers the reasons technology is needed to provide the living conditions taken for granted on Earth—air pressure and composition, temperature control, acceleration, weightlessness ("zero gravity"), nutrition, sanitation, recreation, health, and medicine, to name a few. In a way, it's a survival manual for outer space that also tells its readers what to do if an emergency results from a reduction or cessation of any of the basic living conditions.

Whether the reader is planning to spend a few days in space as a tourist, a longer time as a space worker, or even a lifetime as an emigrant to a space colony, this book can be a useful primer. It is also a reference book to be consulted from time to time while living in space itself. The book is nearly timeless because it's about *Homo sapiens* Mark One Mod One Version 1.0, an organism that hasn't changed for about a million years or so.

The book will point out what we already know about living in space, the type and magnitude of the various hazards and risks involved, and what can be done to reduce these. People don't have to "re-invent the wheel" by making

1

the same deadly mistakes as the pioneers who conquered the air and took the first steps into space.

People will face physical, psychological, and social problems because the environment of space is different from anything encountered on Earth. People cannot live in the space environment itself. They must take into space a small piece of the familiar and comfortable terrestrial environment as they have done when they've journeyed deep beneath the Earth's oceans and high into the Earth's atmosphere.

It's fortunate that this legacy of experience from analogous situations on Earth is available to help solve the problems of space living. In most cases, known or anticipated technologies can be used. However, in these early years of space activities, all the problems aren't solved and all the potential difficulties aren't yet known or understood. This doesn't mean that people shouldn't go into space; it means that they must apply their hands and minds toward solving the problems they encounter. For the first few decades of space living, people who are pioneers will discover these.

Do not forget the warning of the late John W. Campbell, Jr., who advised, "Pioneering involves discovering new and more horrible ways to die." Living in space involves risks, but humans have evolved on Planet Earth by taking risks, sometimes dying but usually winning by learning and adapting.

This book also will help prospective space travelers explain their plans or activities to other people who don't understand why they want to live and work in space instead of staying on Earth where a lot of good jobs are waiting for qualified people. It will help convince those cautious homebodies that outer space is something more than a place to squander all that money that could be used to help people on Earth.

Not everyone wants to go into space or even to see other people do it. Many people are afraid of space because it's the unknown. They're more comfortable working with things as they are. Don't forget that everyone didn't leave Europe, Africa, or Asia to come to the Americas; some stayed where they were and even tried to stop others from leaving. Even today, most people on Earth never travel more than twenty-five miles from the place where they were born.

No matter when and where pioneers, settlers, or colonists set out from their homelands, they had to overcome the opposition of those who remained at home.

The human race has always been and is still made up of a large percentage of staid people who till the fields and mind the store. However, there is and always has been a small cadre of dissatisfied, curious, footloose pioneers who aren't content with things as they are and believe they can make a better life for

FIG. 1-2: *Planet Earth has been the cradle of the human race, but people always leave the cradle to do things in the great universe beyond the nursery.* (NASA photo)

themselves. So they set off toward new frontiers, both physical and mental. The lives of those who stayed at home have been altered beyond imagination by the development of new frontiers. It will be no different in the years and centuries to come.

The restless people are different. Although they may be afraid, they are willing to face the new and the unknown. They do this for any number of personal reasons. They are willing to bet their lives on their ability to make the future better than the past.

For decades to come, the people who live and work in space will be no different from their predecessors who left ancient homelands to venture into the wilderness. Initially, their numbers will be small. But what they see, learn, and do in space—even as space tourists—will have a profound effect upon the lives of everyone else. They will have to exercise every bit of intelligence, know-how, and mental agility they possess. Every branch of human knowledge is involved.

Some people don't know *why* they want to do it, but most of them have a personal philosophy about it that amounts to a way of thinking about themselves and the rest of the universe. A person who does know *why* stands a better chance of succeeding or surviving.

Great achievements in space have indeed occurred without a long-term operating philosophy, only to fail once initial success was attained. The Apollo manned lunar landing program of the United States of America is an example. It was justified on the basis of national prestige and national security. The USA was involved in a cold war with the Soviet Union; as such, the space race and the Apollo program were part of a technological war. Americans could not abide the idea that a bunch of Russians who couldn't keep tractors running could nonetheless travel into space and boast that they would be the first to land on the Moon. The Americans won and the Soviets failed. Beyond landing men on the Moon and returning them safely to Earth, the Apollo program had no clearly defined long-term philosophy and therefore no long-term goals that were understood by those who paid the bill, the American taxpayers. Therefore, that grand project died once it had obtained its limited objective. As a NASA engineer remarked a quarter of a century later, "The Apollo program is finished to nineteen decimal places, but we don't seem to realize it yet."

Today, people want to go into space to live and work, not because of political ideology but because they're doing something that human beings have done for millennia: *using* the universe to produce items of value for themselves and other people. One of the purposes of the human race that we know for certain is that we can and should do this, that we've been successful doing it in the past, and that the universe will let us know in no uncertain terms when we exceed the boundaries of *use* so that it becomes *misuse*. In fact, this has already occurred in several instances here on Earth, and we've learned a lot as a result.

We've used Planet Earth for millennia. Perhaps we've misused it locally because of the greed born of the outmoded philosophy of living in a world of scarcity or because of ignorance of what we were really doing. Some people would even contend that we've misused the Earth itself because we've changed it. However, the Earth's environment and ecosphere have been drastically altered by living organisms ever since life began. Perhaps the purpose of life itself is to change the universe; life itself may be a mechanism for reversing the trend toward disorder, otherwise known as an "increase in entropy." Entropy is defined as the tendency for all energy in the universe to decay slowly to a common level, thus reducing the universe to an overall state of disorder. Life itself may be *antientropic*, creating order out of chaos.

We will probably never know whether or not this is true. The end of the universe—if there is to be an end—may not occur for another ten billion years, depending upon which of the current theories of cosmology one happens to favor. In the meantime, human beings will continue to use what they can.

FIG. 1-3: *The twentieth century began with mankind conquering the air around planet Earth. The most famous photo in aerospace shows the first controlled heavier-than-air flight with Orville Wright at the controls of the* Flyer *as his brother, Wilbur, runs alongside at Kitty Hawk, North Carolina, December 17, 1903.* (NASA photo)

Therefore, if human beings are to use the universe, it means the robot machines sent to the Moon and other planets are just tools to help humans understand those places as a prelude to going there themselves.

The robot machine explorers encountered difficulties that were not anticipated when they were designed. They needed repairs that could not be made because people weren't there to do it. Why were they built and sent to the Moon and planets in the first place?

Basically, they were sent because it was possible to do so using modified military ballistic missiles that are really long-range artillery shells. It was barely possible to use these space-launch vehicles for expensive crewed flight because of their payload limitations; they just couldn't lift the weight of many people and the environmental control systems required to keep them alive in space. In the 1960s, ballistic missiles were cheap to modify and use because their design and production had been paid for by the military services. However, costs kept increasing because extremely high reliability was necessary when they were required to carry people. Furthermore, when the button was pushed and the flight begun, everything had to work perfectly thereafter. Once an artillery shell is fired, it must do its job without any human help. It took forty years for people to understand that spaceships had to be built and operated like airliners, ocean vessels, and road vehicles. People don't like the idea of

FIG. I-4: *The twentieth century ended with people living and working in space. Astronaut James D. van Hoften uses a special tool to work on a scientific satellite before releasing it from the NASA space shuttle.* (NASA photo)

riding around in redesigned artillery shells when a better way to go becomes known.

Furthermore, what the robot explorers discovered posed more questions than answers. Having someone on the spot would have made a lot of difference. Although robot machines can perform in environments that are deadly to people, they can't do what people can do.

A human being can do all that a robot or machine can do and all that it can't, which is 99 percent of everything.

People are absolutely necessary to expand and exploit frontiers.

Can you imagine what history would have been like if a special commission established by King Ferdinand of Spain to evaluate the proposal of the Italian navigator, Cristóbal Colón, to sail westward across the great ocean to the Orient concluded in its final report, "It is far too risky to send sailors on such a hazardous voyage, provided sailors could be hired who are willing to undertake such a dangerous and unknown journey. Therefore, this Royal Commission recommends the development of a crewless exploration ship with a sample-return capability. This vessel would sail westward under automatic control. If it encountered land, it would gather a sample of the soil and reverse its course for the return to Spain."

In this scenario, our ancestors might never have left Europe. Or the exploration of the New World would have been delayed by one or several centuries.

Sending robotic surrogates is not the way human beings really go about pursuing hazardous activities. Humans are risk-taking gamblers, even with their own lives. They race autos and airplanes; they jump out of airplanes; they climb hazardous mountains for fun; they sail around the world in dinghies; they are even willing to pay for adventure. Humans are a crisis-oriented, risk-taking species that thrives not on reason and logic but on hunches and illogic.

The twentieth century has led people to acquire more and more answers to the question, "What is a human being?"

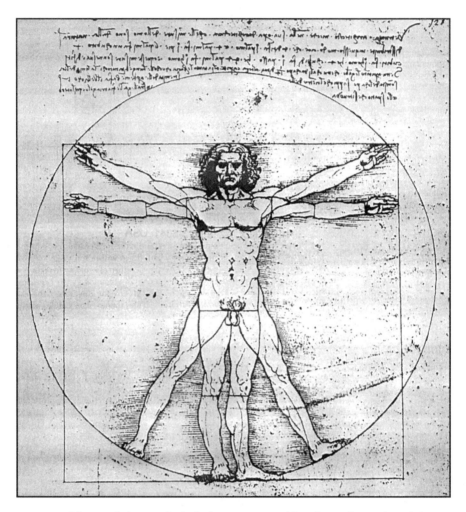

FIG. 2-1: *A human being can be looked upon as a machine, but to live and work in space, people must have an accurate concept of what a human being really is.* Homo sapiens *Mark I Mod 0 Version 1.0 hasn't changed in a long time.* (Leonardo da Vinci)

CHAPTER TWO

WHAT IS A HUMAN BEING?

We MUST NEVER LOSE SIGHT of the fact that those who travel into space or live and work in space are human beings. They may think somewhat differently because they are pioneering a new frontier. However, physically they are no different from people on Earth. Their basic psychological makeup is also the same, because, as have everyone else's, it has evolved over millions of years of evolution.

Therefore, to travel and inhabit space, one must not only know the physical characteristics of the space environment but also something about the human beings involved. A space habitat designed, built, and operated for dolphins, chimpanzees, or goldfish would be quite different in most respects from one designed for people.

Therefore, it's important to know the answer to the basic question: What is a human being?

Why Define a Human Being?

This may appear to be a sophomoric or even impractical exercise. After all, we are all human beings and we know who we are, right? And we should know what we are, right?

Yes, but not quite yes.

For far too long, human beings have conducted their lives and affairs based on folk legends, fairy stories, old wives' tales, and perhaps an overinflated egotistical notion of who and what a human being is.

Progress in the human sciences—not the humanities—has been truly outstanding in the twentieth century. This should come as no surprise to readers who have been trying to keep up with progress in anthropology, biochemistry, neurology, psychology, control mechanisms for complex systems, semiotic analysis, chaos theory, and to some extent the slowly emerging proto-science of sociology.

The Schwartzberg Test

This may raise objections from some people who believe that sociology is a real science. Therefore, let's see if we can define what is meant by "science," and then see what areas of expertise qualify.

Science is universally defined as an organized body of knowledge.

The late Harry L. Schwartzberg of RCA observed that "the validity of a science is its ability to predict." This is based on the assumption that if past performance is known and enough experiments have been conducted with a reasonably large statistical universe, then the future reactions of a system to a given input can be accurately predicted within a known spread of statistical variation. As we grow to understand them better, we find that this holds true even with chaotic systems in which many valid results are potentially viable. In brief, the outcome prediction may include several possibilities, each of which is statistically probable.

In the physical sciences, the Schwartzberg Test can be applied with few caveats except in the newer fields of quantum mechanics, for example, where one of many results are possible. In large and complex systems such as those of interest to the biological scientists, more caveats must be postulated. However, basically at the engineering level, the Schwartzberg Test can be applied. For example, there is a very high probability that the sun will rise at New York City tomorrow morning at precisely the time given in *The American Ephemeris and Nautical Almanac*. There may be a 50 percent probability of rain in San Diego, depending upon how well the meteorologists have guessed that the weather patterns will actually follow the same course as their computer simulations and as they have in the past. (There is a 72 percent chance that the weather tomorrow will be the same as it is today, for example.) There is a high probability that Joe Smith at the corner bar will become intoxicated if in one hour's time he consumes six ounces of a ninety-proof adult beverage, no matter how he brags about being able to "hold his liquor." A public opinion poll conducted with 1,250 people in a selected random sample will produce results that can be duplicated within three percentage points if the poll were taken of the entire population.

These are examples of scientific predictions having varying degrees of validity.

Scientific Measuring Sticks

In 1962, Dr. William O. Davis wrote, "Fundamentally, science must be a series of successive approximations to reality."

However, in any area of science—and particularly in the human sciences—the ability to predict the effect of a given cause can be distorted by the philosophical, ideological, political, or personal beliefs or whims of the scientists involved. It is more difficult for these human factors to enter into the prediction if reasonably precise and accurate measurements can be and are made. Although *no* measurement is 100 percent accurate, it can be accurate enough for practical use. A lot of buildings are put together very well with a carpenter's rule, but that same measuring tool is useless to a micro-machinist or to an astronomer.

To achieve validity, measurements must be made. Alfred, Lord Kelvin, the famous British scientist, put it this way in 1886:

> *I often say that when you can measure something and express it in numbers, you know something about it. But when you cannot measure it, when you cannot express it in numbers, your knowledge is of a meager and unsatisfactory kind; it may be the beginning of knowledge, but you have scarcely, in your thoughts, progressed to the level of science, regardless of what the matter may be.*

Therefore, although many human sciences are still a bit "fuzzy," and even though the sociologists have just started to make some measurements, we still have enough data in hand at this time to make a reasonably accurate and valid definition of what a human being is so that we can properly ensure human survival in the new environment of space.

Defining a Human Being

Any definition of human beings has to be a generalized one because every human being is slightly different. An enormous variation of physical and psychological characteristics is possible within the genetic structure of the forty-six paired chromosomes of an individual's genome as well as within the extremely wide variety of environmental factors to which an individual can be exposed that shape that individual's development.

However, a generalized definition can be stated. It will undoubtedly be modified and expanded as the years go by, just as it has in the past.

A human being is a highly evolved bipedal, bisexual, unspecialized mammalian vertebrate with six unique special features, some of which are shared with other species of apes and monkeys, but whose combination in *Homo sapiens* is unique among all animals:

These six features are:

- an erect posture
- freely movable arms and hands
- sharply focusing stereoscopic color-discriminating eyes
- a large brain capable of judgement, fine perception, and excellent pattern recognition
- the power of speech
- less physiological and psychological differentiation between the sexes than in other species.

Each of these requires elaboration if a human being is going to be properly provided for in space and eventually on Earth itself.

The Erect Posture

The erect posture is unique to human beings and was probably a survival asset that permitted early human-apes to gain the "high ground" of observation when looking over the broad savannahs that became their hunting areas in prehistoric times.

But the evolution of the current human skeleton from that of the four-footed mammals is probably as much of a botched engineering job as is the mammalian, reptilian, or amphibian skeletal structure. Vertebrate skeletal structure was originally evolved for water animals who exist in a world of balanced equilibrium between gravitational and buoyant forces. Gravitational forces affect the internal organs of water animals, but the skeleton merely provides anchor points for muscles and tendons, not a structure that supports the entire mass of the animal against gravity.

Mammalian skeletal structure is therefore full of weaknesses, some of which are evident in human beings.

The vertebrate backbone is not a good structure from which to suspend even a four-footed animal. Humans are not the only mammalian species that has back problems; they are shared with horses and dogs, too. The human backbone probably does a better job supporting the body as a column than a dog's back does as a beam supporting the canine abdomen.

The ball-and-socket hip joint of mammals—not just humans—operates under severely high strains because the forces in the hipbone must be carried around a 90-degree turn just below the ball joint. Dogs suffer from hip problems just as human beings do.

The human rib cage is not as well designed for support as it is in horizontal animals. An extremely complex and unsatisfactory network of muscles is necessary to keep the viscera in place even under normal gravitational forces.

However, the survival value of being able to stand erect overcame the shortcomings of the human skeletal structure. Posture, in common with the other six unique features of a human being, is intimately related with evolutionary survival.

Movable Limbs

One great advantage of the human erect posture lies in the freedom of movement afforded the forelimbs and hands. Although apes may occasionally stand erect, they don't move in an erect manner. Human beings share only with the birds the fact that forelimbs have evolved into different forms that are used differently from the hind limbs.

Because human beings stand erect, the human foot differs greatly from the hind feet of other animals, including the apes. The human foot has a big toe that points forward for walking, not sideward for grasping; the human foot cannot grasp as can an ape's foot.

In common with the birds, human beings must teach their young to move. In all other animals, the knowledge of how to move is preprogrammed into the brain and nervous system before birth. But birds must teach the chicks to fly, and humans must teach their children how to walk upright.

Walking erect freed human arms and hands from locomotion and permitted the further development of the human hand for manipulation. The human hand itself is unique among terrestrial animals. Although many apes possess an opposable thumb, none possess the ability to move each individual finger separately, and none possess the human wrist joint system.

The human wrist may be bent so that the hand is at right angles to the forearm. In this position, the hand may be rotated almost 180 degrees. While doing this (and in any position in between), each individual finger may be moved. When the forearm is held horizontally, the palm of the hand may be turned up or down. No other animal on Earth has this sort of arm or wrist joint.

The human hands and arms may be rotated in two nearly overlapping hemispheres from the shoulder, providing a human being with a large area about the body in which the hands and fingers can manipulate objects.

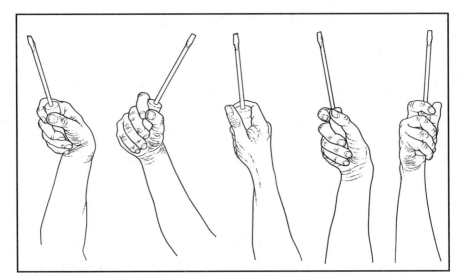

FIG. 2-2: *The human wrist is a marvelous device with degrees of freedom in movement not possessed by any other animal.* (Art by Sternbach)

The tip of each finger is a dense mass of tactile and kinesthetic receptors, providing a human being with a sense of touch perhaps unrivaled in the animal kingdom, with the possible exception of the dolphins.

A large portion of the human brain is devoted to the task of sensing and controlling the human hand. More brain volume is dedicated to the thumb than to the entire visceral cavity.

A human being has two outstanding manipulators that are unmatched in their ability to feel the surrounding environment and to handle tools. Although humans may not have as acute senses of hearing and smell as other animal species, human beings' sense of touch is unparalelled in the animals kingdom.

Sharp-Focusing Stereoscopic Eyes

In spite of the fact that some bird species are believed to have more acute visions, one of the unique human features is stereoscopic, sharp-focusing, color-discriminating eyes.

Like the apes, a human has two eyes set in front of the skull with overlapping fields of vision. Some other mammalian species have eyes set partly on the sides of the skull while amphibians and fish have eyes with no overlapping visual fields whatsoever. This overlapping of visual fields permits stereo vision which gives humans outstanding depth perception while manipulating tools or other materials.

The human eye also possesses remarkable ability to "accommodate," or change the shape of its lens, to focus on objects closely held in the hand.

The retina, or visual receptor of the human eye, excells at color discrimination. The retina contains color-sensitive visual receptors known as cones (from their general shape) which are tightly packed in the retina around the prime focus of the eye system. The majority of the retina is made up of rod-shaped visual sensors that are not sensitive to color but are instead far more sensitive to light. At high-light levels, cone vision overwhelms rod vision, and humans see in color and in great detail; at low light levels, rod vision predominates, color washes out to various shades of gray, and the eye becomes primarily sensitive to motion. Therefore, although color detail is sensed near the primary axis of vision, the remainder of the eye's visual field is primarily used to detect motion. Thus, the human visual system performs the dual survival roles of being able to sense in detail and in color whether or not the fruit held in the hand is indeed ripe while the peripheral visual field is constantly alert to sense the motion of something lurking in the bushes.

The Large, Complex Brain

The factors of erect posture, highly sensitive and manipulative hands, and sharp-focusing binocular color vision required the development of a large brain in terms of brain weight per unit of body weight. Furthermore, they required the evolution of specialized parts of the brain.

Standing erect on only two legs is exceedingly difficult because it's an unstable condition. An enormous volume of brain is required to handle the constant input from tactile sensors in the feet and legs, the sensory input from the balance organs located behind the ears, and visual cues from the environment. Human children will pull themselves erect as apes do, but only human children can be taught to walk upright. They must be *taught* because the ability to walk is not an inherited characteristic. This requires the use of a large amount of brain volume devoted to perception and association.

The ability to use the hands and fingers requires similar brain power. Human children spend a great deal of time learning how to use their hands. Again, they must be taught to manipulate an eating utensil, which is basically the first tool a child must master. As the human child grows and matures, the manipulation of other tools must be taught.

The ability to play the piano, paint a picture, or use a screwdriver must be taught. These abilities require a brain that possesses the additional feature of

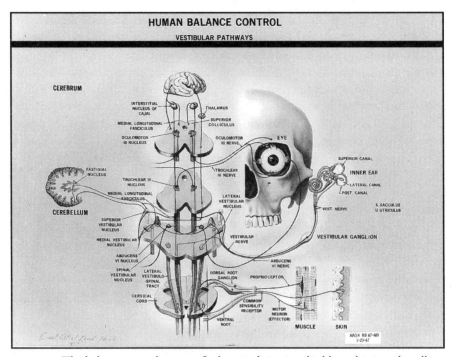

FIG. 2-3: *The balance control system of a human being is a highly sophisticated, well-evolved system of organic sensors, colloidal computers, and output peripherals.* (NASA photo)

being able to exercise judgment and make decisions, abilities which must also be taught. Some recent work in psychology indicates that the left side of the human brain is dominant in learning these skills and in controlling the pertinent parts of the human body to manipulate the outside world, while the right side of the brain has the ability to use judgment and discretion.

These four unique features require a brain that is large in relation to the rest of the body. A newborn human child is almost a brain with a small body attached. And, unlike other mammalian species, that big brain must be taught if the child is to become a human being.

The Power of Speech

The need to be taught required a fifth unique human factor, the power of speech. Other apes possess a limited power to communicate by sounds. Gibbons have a vocabulary of nine separate sounds consisting of imperatives such as, "Stop! Danger!" or "Stay away from my mate!" Animals in general can verbally warn or command, but they cannot verbally transmit past experiences or future plans to others of their species. In this regard, the power of speech makes humans unique.

Language probably developed out of the need to teach human children what to do—to walk, to hunt, to make tools, to use tools, etc. The human uniqueness in locomotion and transport, perception, and the ability to handle tools and materials is not preprogrammed into the human brain as less developed factors are in animals. A kitten knows how to clean itself, climb trees, and hunt rodents. The human brain evolved from such a preprogrammed, instinctually driven control organ into one that is programmed to *learn*. The power of speech and the development of language evolved from this. The power of deductive logic and the act of creation grew from the power of speech because human beings think in words.

When someone studies another language, they should also study the culture in which the language is used. Every human language is a *complete* communications system that mirrors as well as satisfies the cultural needs of those who speak it. It also mirrors the cultural way of thinking. For example, in a culture where little change takes place, the structure and grammar of the language becomes complex, formalized, and stiff. As a culture becomes more complex and undergoes rapid changes, the structure of the language grows simpler to permit rapid lingual change. Many language teachers, writers, lexicographers, and grammarians decry the current world tendency to speak mixed languages where words are borrowed outright from other languages; most of these people also resist the vulgarizing of popular speech and writing. The greatest resistance is mounted against the use of jargon. However, lingual mixing, vulgarization, and jargon/slang are age-old factors that permit a spoken and written language to continue to develop as a means of communication of new concepts, ideas, principles, and techniques.

The human development of a large brain capable of handling the unique developments of erect posture and bipedal locomotion, free-moving hands and arms capable of precise manipulation, fine binocular color vision, and the power of speech for enhanced communication all permitted human beings to achieve mastery of Planet Earth. However, an additional factor has recently been discovered that undoubtedly provides great assistance in doing these things.

Differentiation of the Sexes

Human males and females show less sexual differentiation that any other animal species. That is, there are fewer strong physical and psychological differences between men and women than in other species. Both men and women share numerous secondary sexual characteristics. Both sexes exhibit amazingly similar physical strengths and endurance to a degree not believed possible even a few decades ago. Although the human female must physically bear the human child, either parent is equipped to raise that child to be an adult human being.

This is probably not a new thing in the human race. Once, human beings were a small, rare breed; there may have been fewer than a thousand of them on Earth. They couldn't afford the luxury of the division of labor during the hunt because there were too few humans. When humans began to till the soil, they were still few in number and half their children died of natural causes within five years of birth. Those males and females who could work together in teams survived; those who were weaker than their mates didn't. It was only when human beings conquered all other species and became masters of the planet that their numbers began to increase and they started expanding into every ecological niche they could manage to occupy. Then the division of labor became a necessity. The first division of labor was between male and female.

Because good times were few and scarcity was the general lot of humanity for most of the time on most of the planet, this cultural sexual dimorphism was artificially encouraged. It is still strong in those cultures living in scarcity. In modern advanced cultures where abundance is becoming the general rule, sexual differentiation is disappearing and the original teamwork and equality between human sexes is reappearing.

Whereas the mastery of transportation, communication, and material handling gave human beings mastery of the planet, the unspecialized nature of the sexual team may be the key to the mastery of space because it may be the critical factor in the development of a successful extraterrestrial culture.

The Cultural Animal

The physical development of the human species has of necessity created human beings as cultural creatures. The ability to pass on to their children and others the experience of the past and the potential for the future made possible by the power of speech is itself the root of culture.

The late historians Will and Ariel Durant wrote in the introduction to their monumental series, *The Story of Civilization:*

> *Civilizations are the generation of the racial soul. As family-rearing, and then writing, bound the generations together, handing down the lore of the dying to the young, print and commerce and a thousand ways of communication may bind the*

civilizations together and preserve for future cultures all that is of value to them in our own. Let us, before we die, gather up our heritage, and offer it to our children.

But what are the essential elements of a human culture, a group of individual human beings drawn together and interacting between themselves within the group and in a unified manner with other human groups? What are the environmental and social elements that underlie all successful cultures of the past and must therefore provide the foundations for people living together in space?

Again, the smallest element in any human institution is the human individual, regardless of the ideology of the group.

Modern human cultures appear to be complex because of the many interrelating institutions in which an individual participates. Anthropologists have found that all the basic elements of institutions are present in the simpler hunting cultures from which our modern ones evolved. Therefore, it's possible to get an excellent and highly detailed picture of the cultural and environmental requirements of a human being by referring to these simpler cultures. The extremely close relationship with modern living will immediately become evident.

The Open Environment

Human beings are physically attuned to the specific characteristics of the terrestrial environment. This environment must be taken into space with humans for reasons that will be discussed later in detail. But beyond the physical characteristics of the environment are those which are more subtle and subjective: the psychological and social ones.

The human being evolved as a hunting mammal who is happiest if the cultural environment contains space to move around in. These surroundings should be as "natural" as possible or give the impression of being natural. They should include other animals, if possible, and green growing plants. This is an evident trend in modern post-industrial cultures where animal pets have taken the places of barnyard animals and thriving indoor plant and outdoor gardening and landscaping industries have emerged.

The attempt to achieve this environmental goal exists in all human cultures even if, in some, it is a reality only for a few individuals. Actually, it is a requirement for all mammals. Scientific experimentation has revealed that when too many rats are placed together in too small a space, they begin to exhibit bizarre behavior of a nonsurvival type. Human beings are the only animals to do this to themselves voluntarily. Everywhere such crowding exists, especially in the core of large cities, social problems come into existence that appear to be unsolvable. And, people being what they are, perhaps these problems remain unsolvable as long as the environment remains unchanged.

The Basis for Work and Play

Humans need physical exercise and recreation. One reason most Europeans and Asians have learned how to live together reasonably well and successfully in the restrictive space of cities is that these cities have parks and places for play. Modern biotechnology has rediscovered a physiological and a psychological need for exercise.

The physical and mental problems of people who are, physically and psychologically, hunters who work at tasks in a sedentary environment are now well known but perhaps not yet completely understood.

And human beings need recreation. "Play" is not a unique activity among humans. Animals engage in playful, nonhurtful competition and individual interaction either as practice for similar nonrecreational activities or as diversion.

Play is the foundation for creativity, for the achievement of the willing suspension of disbelief that permits a socially acceptable way of acting out fantasies, and for the impetus that underlies both science and art. This exists in all human cultures.

A human being needs activities that are at times dangerous, that at all times require the use of more brain than brawn, and that sharpen the wits by presenting something new and different that appears to be (or is believed to be) solvable and doable. Hunting is an occupation that includes all of these. Farming doesn't. Nor does work on a factory assembly line, the dehumanizing legacy that has come down from the early days of the First Industrial Revolution in England. Although many people long for a life of leisure, those who have been able to live such a life soon discover that they become bored or ill, either physically or mentally.

Social Interaction

A human being needs a small group of people with whom informal face-to-face interaction can take place without being perceived as a threat or being threatened. This group may include family, close friends, or even a formally organized club. If these groups become too large, the individual feels a loss of control over personal relationships within the group and a threat of possible coercion by the group leaders over which no personal influence can be exerted.

As social groups become large, formalized group behavior develops because it becomes important to know who is in charge of what, what individual responses must be in any given situation, and what responses are expected from others.

Because human beings evolved as superior hunters who could cooperate in the kill of much larger animals, people know that cooperation is absolutely essential when large and dangerous jobs are tackled.

The hunter's mind also wants to make personal experience available to the group to ensure success. Therefore, people want the freedom to express opinions, present data from experience, and form their own opinions based on their evaluation of the situation or the individuals who lead or propose to lead the group.

When individuals don't do what is expected of them or when they perform destructive acts against other people in the group, human beings must indicate the wrongness of such behavior and attempt to prevent its recurrence by inflicting punishment that ranges from *lex talionis*—an eye for an eye—to banishment from the group. An organized system or code of behavior must be developed that includes provisions for handling behavior that deviates from the code. Thus a legal system that embraces the group's concepts of justice emerges. Such systems are most effective when the majority of the members of the group participate in the process, and may become totally ineffective if such power gravitates into the hands of a few.

Usefulness in Space Living

All of this and more is applicable to living and working in space, including the development of a new extraterrestrial civilization. Even in the early, primitive days of its development in the twenty-first century, this hard-won knowledge of who and what human beings are must be applied to what is done in space.

And although in space it is possible to start afresh without the ghosts of the past to haunt the endeavor, it's expensive and deadly to make the same mistakes all over again. It's extremely dangerous to re-invent the wheel, especially in the already dangerous environment of space (which is no more dangerous or terrifying than the opening of other frontiers on Earth). Human beings will have to make use of all their unique characteristics, and especially the ability to communicate the lessons of the past to the generation of the future.

FIG. 3-1: *Mankind's journey into space began when people first left the surface of the Earth in 1783 to travel in the atmosphere using primitive hot-air balloons.*

CHAPTER THREE

ATMOSPHERES

STAYING ALIVE, HEALTHY, AND HAPPY IN SPACE depends upon many technologies. But basically, human life and health in space depend upon Earth-evolved physiology. Before people can work with the nuts and bolts of life-support systems, they need to know what those systems must support in the first place. Some life-support systems are more critical than others. The safety rules, developed from years of experience, depend upon some basic facts about a person's physiology.

The first area of life support that must be approached involves atmospheres.

Basing a discussion of the human problems arising from flight in the Earth's atmosphere may appear at first glance to have little relationship with living in space. However, the conquest of space was preceded by the conquest of the air—the Earth's atmosphere—during which much was learned about the importance of physiological effects of zero pressure, reduced pressure, and even overpressure on the physical body, factors that people living in space must contend with every second.

Without a surrounding atmosphere at adequate pressure and with proper composition, a human being dies—within minutes. Insofar as a human being is concerned, space begins at the Earth's surface itself. Because of the unique characteristics of the Earth's atmosphere, a person can experience conditions approaching those in space without leaving the Earth's atmosphere. As one ascends upward through the Earth's atmosphere, conditions become more spacelike. This led a pioneer space scientist, Dr. Hubertus Strughold, to propose the concept of "partial space equivalency" as a function of altitude above the Earth's surface.

The space between the worlds is devoid of any sensible atmosphere and, from the functional point of view of a human being, can be considered a vacuum.

The gaseous atmosphere of Planet Earth is unique in the solar system. Other planets and large moons possess atmospheres. However, these do not have the necessary combination of proper pressure and composition that will permit a person to live without life-support equipment.

The Earth's Atmosphere

In the eighteenth century, people believed that the Earth's atmosphere was everywhere the same, making it possible to fly to the Moon in an open basket lifted by trained birds or even a balloon. The atmosphere was believed to permeate the entire universe. The personal experiences and reactions of people who lived or traveled in high mountains were ignored or misinterpreted because of this "pervasive atmosphere" belief. It wasn't until scientists began to measure various atmospheric characteristics and explorers started to ascend to high altitudes in balloons that the atmosphere's true nature was discovered.

The first scientist to measure an atmospheric characteristic was Evangelista Torricelli (1608–1647). A colleague of Galileo, he was a mathematician and physicist who lived in Florence, Italy. In 1643, he invented the mercury barometer (see Figure 3-2) and concluded that it was the weight of the atmosphere that sustained the column of mercury inside the inverted closed tube. This was quite contrary to the beliefs of the time based on the doctrine of the ancient Greek philosopher Aristotle who had stated that "nature abhors a vacuum." In 1648 Florin Périer carried a "torricellian tube" up the mountain Puy de Dôme and found that the height of the mercury column was less at the summit.

The pressure of the Earth's atmosphere at sea level will support a column of mercury 760 millimeters (29.92 inches) high. The amount of pressure required to do this is 14.696 pounds per square inch. The pressure is created by the weight of trillions of air molecules in a column one-inch by one-inch extending to the top of the atmosphere. The air molecules are pulled toward the Earth and are thus given weight by the Earth's gravity.

Minor daily and seasonal variations of atmospheric pressure at any point on the Earth's surface are caused by something that the Earth has and space does not: weather. Localized pressure variations of as much as two inches of mercury or one pound per square inch are not uncommon. In the middle of a hurricane, pressure can drop as low as 26 inches of mercury or about 2 pounds per square inch less than normal.

As a person ascends into the atmosphere or climbs a mountain, the column of air grows shorter and thus the pressure exerted by that air column becomes less. Near the surface of the Earth, the atmospheric pressure decreases roughly one inch (25.4 millimeters) of mercury, or about a half-pound per square inch, with every 1,000 feet of altitude.

The Earth's atmosphere is a mixture of gases having approximately the following makeup (numbers rounded off):

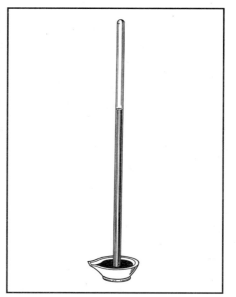

FIG. 3-2: *The mercury barometer or "torricellian tube" in which the column of mercury in the tube is held up by the air pressure on the surface of the mercury in the lower bowl or reservoir.* (Art by Sternbach)

Nitrogen:	78.09%
Oxygen:	20.95%
Argon:	0.93%
Carbon dioxide:	0.03%

Other gases such as neon, helium, krypton, hydrogen, xenon, ozone, and radon make up the minute fraction of a percent remaining.

This composition doesn't vary on a worldwide basis except in the vicinity of large cities and industrial complexes where more carbon dioxide and particulate matter may be concentrated. However, there is a greater percentage of carbon dioxide over the oceans of the southern hemisphere than over any large industrial complex.

Although pressure decreases as altitude increases, the composition of the atmosphere remains reasonably the same below 50,000 feet or 15 kilometers.

The density of the atmosphere—i.e., its weight per unit volume—under Standard Temperature and Pressure (STP) conditions of 59 degrees Fahrenheit and 14.696 pounds per square inch is 0.76475 pounds (mass) per cubic foot. As altitude increases, air density decreases along with pressure. This means that at high altitudes there are fewer molecules in a given volume of air.

The "sea level" or normal characteristics of the atmosphere are important to know because (a) they represent the conditions that the human body con-

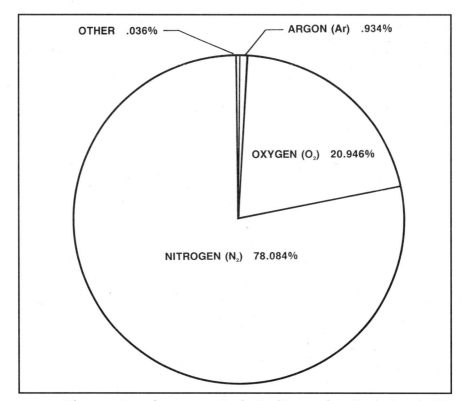

FIG. 3-3: *The proportions of various gases in the Earth's atmosphere.* (Art by Sternbach)

siders "normal," and (b) they are the conditions that will have to be maintained around human beings in space.

Oxygen

A human being cannot stay alive for more than a few seconds without a source of oxygen at a minimum gas pressure. Because the human body has evolved in the unique gaseous atmosphere of Planet Earth and uses the oxygen component to support life, the pressure and composition of an atmosphere available in a space module or suit must closely resemble sea-level conditions on Earth.

This critical environmental factor should be a verity that is well understood by everyone. However, it's taken for granted and generally misunderstood. Until only recently in biological time, humans have always lived, traveled, and worked with the Earth's atmosphere around them. Only in the past two centuries could people leave the surface of the Earth and travel in its atmosphere.

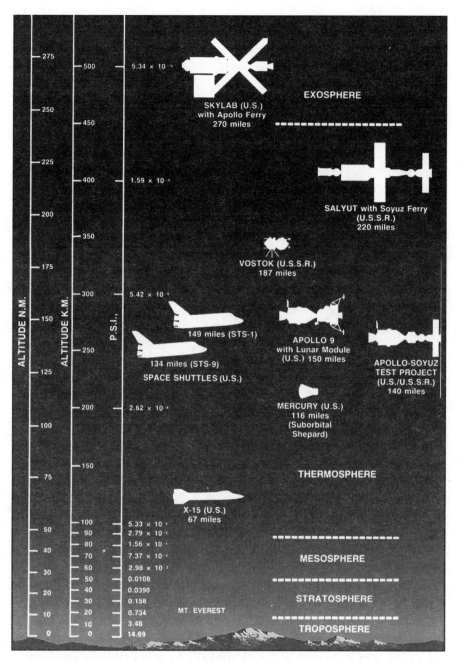

FIG. 3-4: *Going to and from space, spaceships and their passengers must pass through the Earth's atmosphere whose regions have been divided into layers by scientists. Human beings evolved at the bottom of this ocean of air.* (Art by Sternbach)

And only within the last two hundred years have humans learned the primary purpose of breathing.

The fact that the Earth's atmosphere changes with altitude provided the first clue concerning how the human body is affected by surrounding atmospheric pressure and how strongly dependent upon it is the operation of the human body. Dr. Strughold's "partial space equivalency" was experienced by people within a few years after the French Molgolfier brothers made their initial manned balloon flights near Paris in 1783.

The human body is a heat engine that consumes the fuels of carbohydrates, fats, and proteins from food by the chemical process of oxidation, which requires the presence of oxygen. Oxygen is obtained from the surrounding atmosphere by the process of respiration using the lungs. The true function of the human lungs was discovered in 1660 by Marcello Malpighi (1628–1694) in Bologna, Italy. The fact that the lungs provide the human body with oxygen and remove the gaseous products of cellular oxidation—carbon dioxide and water—was discovered by the French physiologist Clause Bernard (1813–1878).

A human being at rest consumes approximately a half-pint of oxygen gas per minute at standard temperature and pressure (STP) but can demand as much as 8 pints per minute while running.

Inhaled oxygen molecules from the air must pass through the alveolar membranes of the human lung into the capillaries that carry blood through the lungs. Then the oxygen molecules must attach themselves chemically to the red blood corpuscles. To do this, they must have enough energy in the form of pressure to dislodge the carbon dioxide molecules carried by the red blood cells from other parts of the body.

Pressure

John Dalton (1766–1844) discovered that, in a mixture of gases, each component gas behaves as if it alone occupied the total volume of the gas mixture and therefore exerts a gas pressure equivalent to its percentage in the mixture. Known as Dalton's Law of Partial Pressures, this can be restated: "The sum of the partial pressures of each constituent gas in a mixture equals the total pressure of the mixture."

And it was the French physiologist Paul Bert (1833–1886) who pointed out in 1878 that it's the partial pressure that determines the physiological effects of a gas.

Thus a gas mixture such as air containing 20 percent oxygen with a total pressure of, say, 10 pounds per square inch (psi) has an oxygen partial pressure of $10 \leftrightarrow 0.20$ or 2.0 psi.

When applied to the Earth's atmosphere at STP, Dalton's Law states that human beings are used to living in an atmosphere of oxygen with a pressure of 3.08 psi.

Pressure measurements made inside human lungs have shown that the pressure of carbon dioxide on the blood side of the alveolar membrane is approximately 40 millimeters of mercury or 0.7735 psi. This is about one-fourth as much as the oxygen partial pressure in the Earth's atmosphere at STP. As long as the oxygen partial pressure in the lungs is greater than the partial pressure of carbon dioxide, oxygen will flow inward across the membrane to replace the carbon dioxide, and the carbon dioxide will flow outward so it can be expelled from the body with the exhaled breath.

These physical facts provide the foundation for the most critical and hazardous danger to humans who fly airplanes at high altitudes or who live in space: hypoxia, anoxia, and hyperventilation.

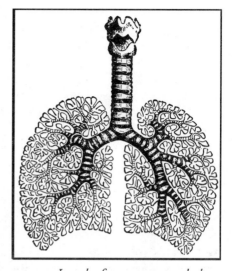

FIG. 3-5: *In order for oxygen to reach the bloodstream and then be transported to body tissues where it's needed for metabolism, oxygen molecules must have enough energy to cross the membranes of the alveolar sacs in the lungs and then replace the carbon dioxide being carried by the red blood cells.*

Lack of Oxygen

Hypoxia is a condition of oxygen shortage in the human body while *anoxia* is a condition where insufficient oxygen is available to sustain life. They are two aspects of the same environmental factor and differ only in degree. The condition of hypoxia can exist for hours or more while anoxia can be fatal in a matter of minutes. Because of the importance of pressure, hypoxia and anoxia cannot be considered apart from *hypobaria* (reduced pressure) and *abaria* (lack of pressure).

Aviators flying at high altitudes in the atmosphere were the first to experience hypoxia. Hypoxic situations were reported during the First World War when the fragile airplanes of the time began to achieve altitudes in excess of 15,000 feet above sea level.

The cause of hypoxia is related to partial pressure. As altitude increases, the partial pressure of oxygen decreases. Air density also decreases, meaning that fewer oxygen molecules occupy a given volume of air. The combination of lower oxygen partial pressure and lower air density means that when a human inhales surrounding or "ambient" air, fewer oxygen molecules are available to pass through the alveolar membrane and into the bloodstream. Furthermore, because of the reduced partial pressure, oxygen molecules enter the lungs with less available energy.

Hypoxia, whether encountered while flying or in space, is perhaps the most insidious of all dangers because the hypoxia victim doesn't realize it's happening.

The first portion of the body to suffer from oxygen deprivation is the most important one: the brain. Although the human brain comprises only about two percent of a human's body weight, it requires 25 percent of the oxygen inhaled during respiration. Within the brain, the "higher functions" or the most recent evolutionary portions are the parts initially affected by hypoxia.

The first symptoms of hypoxia are insidiously pleasant and resemble mild alcohol intoxication:

- Normal self-critical ability is dulled.
- The capability to exercise judgment by comparing and analyzing alternatives is greatly impaired.
- The brain's ability to correlate different sensory inputs disappears.
- Memory becomes elusive.
- Important matters no longer seem important because hypoxia has a tranquilizing effect.

Motor control and coordination suffer next. A person becomes clumsy without being aware of it. Hypoxia brings on feelings of well-being, drowsiness, nonchalance, and a false sense of security.

The last thing a person believes to be necessary is additional oxygen.

At night or under conditions of reduced illumination, vision suffers and becomes dim, again without the person being aware that it's happening.

Many of the same hypoxic sensations are experienced regularly and voluntarily by people who smoke tobacco products. Nicotine has a much greater affinity for red blood corpuscles than oxygen. Thus, a smoker deliberately subjects the brain and body to a mild form of hypoxia as well as a carbon dioxide imbalance. Airplane pilots who smoke are known to have a lower tolerance to hypoxia than nonsmokers. By smoking a single cigarette, a smoker automatically puts his brain at an altitude of 8,000 to 9,000 feet because of the carbon monoxide in the inhaled smoke.

26,000 Feet — Chamber Test

INSTRUCTIONS:

Log Distances and Leg Times were given and Pilot was asked to Compute Ground Speeds and Total Times.

Oxygen Mask Removed after 4th Entry. Note the errors and Deterioration of Legibility.

1 Misread Computer

2 Used Wrong Computer Scales

3 Error in Addition

4 Number Omitted

5 Loss of Consciousness

	GROUND SPEED	DISTANCE N.M.	TIME	
			LEG	TOTAL
	153	28	11	11
	180	63	21	32
	174	110	38	70
	180	36	12	82
1	174	43	15	107 (3)
2	212	76	28	35 (4)
5	\	80	29	

FIG. 3-6: *The effects of hypoxia are shown when a pilot in a pressure chamber was asked to make and write down calculations when the chamber was at various equivalent altitudes. As altitude increased, the mistakes became more obvious until unconsciousness set in. (Federal Aviation Administration)*

A person who smokes is at high risk in space not only because of these factors but also because of the effect of tobacco smoke tars and other ingredients upon components of the life-support systems. Before smoking was banned on domestic airliner flights in the United States, a quick visit to any local airline maintenance facility would have revealed the brown goo that had to be removed regularly from fans, blowers, filters, and other parts of the cabin air-conditioning system of a jet airliner. In space facilities, this deliberate gumming up of the life-support system cannot be tolerated because if the system stops working, so do the people who depend upon that system.

As hypoxia gets worse, a person loses the ability to balance and gets dizzy. There may be a tingling of the skin. A dull headache may begin, but it's usually only partly perceived at this stage and is dismissed as a mild annoyance because of the advanced state of tranquilization. The heart rate increases as the heart pumps more blood through the body. The lips and the skin under the fingernails turn blue. The field of vision narrows until peripheral or cone vision disappears, leaving only rod vision that itself slowly becomes blurred as nerves that control the ability to accommodate or focus begin to stop working because of lack of oxygen.

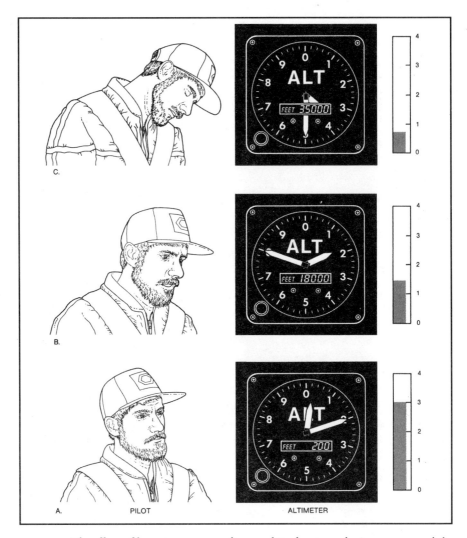

FIG. 3-7: *The effects of hypoxia vary according to altitude, atmospheric pressure, and the partial pressure of oxygen.* (Art by Sternbach)

Unconsciousness finally occurs if the oxygen partial pressure continues to decrease or drops to a level where insufficient oxygen is available to the brain.

Anoxia

At that point, the condition of anoxia is predominant. The person may go into shock or convulsions caused by carbon dioxide poisoning that upsets the acid-base balance of the body chemistry, creating acidosis. This condition alone may kill the person long before the oxygen level in the body drops to a point where the brain stops working altogether.

Preventing Hypoxia and Anoxia

The first data concerning hypoxia came from aviators. When people fly at high altitudes in unpressurized cabins, safety regulations now require the use of on-board supplies of supplementary oxygen. Gas from bottled oxygen is usually fed to the crew and passengers via oxygen masks they place over their noses and mouths. Supplemental oxygen should be used at all times when flying higher than 12,500 feet. At that altitude, the partial pressure of oxygen is 1.92 psi or about 62 percent of its sea-level value.

However, hypoxia can begin at altitudes as low as 8,000 feet (oxygen partial pressure of about 2.3 psi or 75 percent of that at sea level) with people who are overweight, elderly, or heavy smokers. Since hypoxia drastically affects cone vision, which is essential for night vision, pilots at night often begin to use supplemental on-board oxygen at altitudes of 8,000 feet.

Thus, the lower limits for the onset of hypoxic effects can be set at an oxygen partial pressure of about 2.0 psi.

Severe hypoxia occurs at 18,000 feet of altitude where the total ambient pressure and therefore the oxygen partial pressure is half that at sea level (see Figure 3-7). At an altitude of 18,000 feet, a person is physiologically halfway to space. Half the atmosphere is below 18,000 feet and the partial pressure of oxygen has dropped to 1.54 psi.

It is possible to breathe supplemental oxygen through a face mask up to an altitude of approximately 35,000 feet where the oxygen partial pressure is 0.7735 psi. This is equal to the carbon dioxide partial pressure in the lungs. Thus no energy or pressure differential exists to allow the oxygen to replace the carbon dioxide in the blood in the lungs. Therefore, oxygen must be supplied to the lungs at a pressure greater than ambient. This is called "pressure breathing." Oxygen under pressure inflates the lungs and can displace the carbon dioxide. However, the person must forcibly exhale against the pressure. This is an opposite action to normal breathing and becomes extremely tiring after a short period of time.

At altitudes in excess of 45,000 feet, the pressure differential between the oxygen in the pressure-breathing system and the total ambient pressure is too great, and it becomes impossible for a person to exhale.

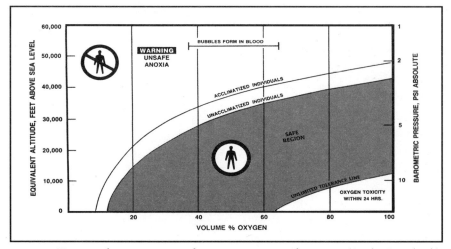

FIG. 3-8: *Human tolerances to atmospheric composition and pressures.* (Art by Sternbach)

Abaria

Pressure breathing with an oxygen mask cannot be used above 45,000 feet or in outer space where the ambient atmospheric pressure is zero. Pressure must be applied to a person's entire body either by surrounding it with an artificial atmosphere or by applying physical pressure to the entire body surface.

Besides oxygen pressure, boiling points are affected by atmospheric pressure, as well. As atmospheric pressure decreases, the boiling temperature of water (and other liquids) decreases. Anyone who has tried to cook a meal at high altitudes in the Rocky Mountains is aware that boiling water is not as hot as it would be at sea level. At an elevation of 18,000 feet where the ambient pressure is 7.32 psi, water boils at 180 degrees Fahrenheit (82.2 degrees Celsius). At an altitude of approximately 65,000 feet where the ambient pressure is about 1 psi, the boiling point of water is depressed to 98.6 degrees F (37 degrees C), which is blood temperature. Technically, without the application of additional pressure by a pressure suit or cabin, blood will boil inside the human body. In actuality, this doesn't happen because of the skin, as will be discussed later.

Hyperventilation

It's as hazardous for a human to get too much oxygen as it is to get too little. Rapid, deep breathing called "hyperventilation" can occur voluntarily but usually takes place under stress. It's easy to experience a mild form of hyperventilation, and it's a common children's game. Just begin breathing rapidly

Partial Pressures of Atmospheric Gases as a Function of Ambient Pressure				
Gas	Percentage of Air	Ambient Pressure (psi)		
		14.7 (sea level)	7.35 (18,000')	3.46 (35,000')
Nitrogen	78.09%	11.476 psi	5.738 psi	2.711 psi
Oxygen	20.95%	3.079 psi	1.539 psi	0.727 psi
Carbon dioxide	0.03%	0.004 psi	0.002 psi	0.001 psi

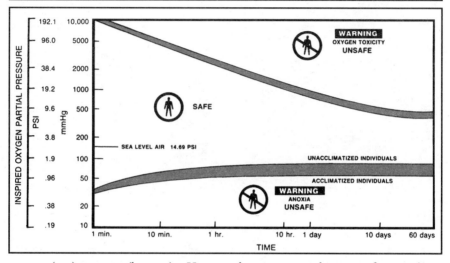

FIG 3-9. (top). FIG 3-10 (bottom). *Human tolerance to partial pressure of oxygen.* (Art by Sternbach)

and deeply. The symptoms of hyperventilation will become apparent very quickly. The age-old survival mechanism of our hunter ancestors goes into action, pouring the substance adrenaline into the bloodstream and triggering the human body into an emergency condition that includes rapid, deep breathing in anticipation of immediate and extensive action.

However, hyperventilation can flush too much carbon dioxide out of the blood and lead to oxygen toxicity. The presence of carbon dioxide in the blood causes the blood serum to be slightly acidic. Therefore, removing too much carbon dioxide from blood changes its chemical balance. This produces dizziness, tingling of the fingers and toes, a sensation of body heat, increased heart rate, blurred vision, muscle spasms, and, if carried out long enough, unconsciousness. The symptoms are very similar to those of hypoxia although hyperventilation is brought on by precisely the opposite conditions.

Unlike hypoxia and anoxia, hyperventilation can be overcome quite easily

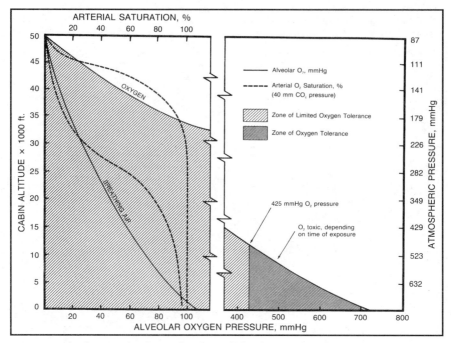

FIG. 3-11: *Blood oxygen levels as a function of alveolar pressure of oxygen.* (Art by Sternbach)

once the condition is recognized. The affected person must be certain that it's indeed hyperventilation and not hypoxia brought about by a failure of the life-support system, the pressure suit, or the pressure cabin. If everything appears to be normal, the condition can be verified as hyperventilation. A person must then slow down the breathing rate as well as the depth of each breath. The easiest way to do this it to talk, sing, or count aloud. Normally paced conversation tends to slow and stabilize breathing rate, and so does singing. Normal breathing is an immediate cure for hyperventilation.

Nitrogen and the Bends

Deep-sea divers and free divers using scuba equipment have encountered the opposite conditions from outer space because they continually work under *increased* environmental pressure depending upon the depth to which they dive. They've encountered a phenomenon known as nitrogen narcosis or "rapture of the deep." Although people living in space may not encounter this because it occurs under pressures of 50 psi or more, other problems may be encountered with nitrogen.

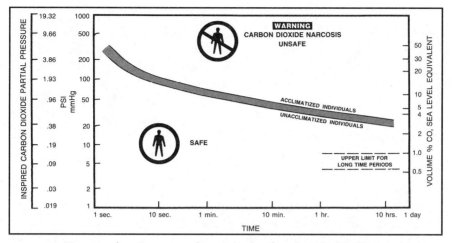

FIG. 3-12: *Human tolerance to partial pressure of carbon dioxide.* (Art by Sternbach)

As with most other environmental factors, either too much or too little nitrogen causes trouble. For example, shortly before the second flight of the NASA space shuttle in 1981, a launchpad worker was killed because he unknowingly entered a compartment in the tail of the Orbiter that was filled with nitrogen used to flush other toxic gases from that space. Exposure to a pure nitrogen atmosphere is almost instantly fatal although nitrogen is a harmless, nontoxic gas that makes up more than 70 percent of the normal atmosphere people breathe. Several lungsfull of nitrogen displace all the oxygen there, leaving the blood loaded with carbon dioxide. Death is rapid and painless.

Nitrogen is the preferred diluting gas for all space facilities because it's the major diluent gas in the Earth's atmosphere. Humans are used to it. However, people living in space may encounter imbalanced nitrogen-rich atmospheres, another hazard in space pioneering.

Although nitrogen is preferred over other gases such as helium (long used by deep ocean hard-suit divers) because of the relative availability of nitrogen, the use of nitrogen as the primary diluent creates the potential for other major problems in space, problems that have been known on Earth for decades by divers, underwater caisson workers, and aviators.

This major concern is that of evolved gas, called "nitrogen narcosis" or "the bends."

Like oxygen and carbon dioxide, nitrogen also crosses the alveolar membranes of the lungs and goes into solution in the blood serum. When a person's body is subjected to a sudden loss of pressure, nitrogen dissolved in the blood can come out of solution and form tiny bubbles.

This is the same physical mechanism that takes place when a bottle of soda water, beer, or carbonated soft drink is opened. While the contents of the bottle are under pressure, the carbon dioxide that makes the fizz remains in solution. The instant the pressure is released by opening the bottle, the carbon dioxide comes out of solution as tiny bubbles of gas.

When this happens to the nitrogen in a depressurized human body, the nitrogen gas bubbles tend to congregate in the arm and leg joints where their presence creates almost unendurable pain, "the bends."

However, the bends are rarely observed with changes in pressure of 7 psi or less.

There is also a solution to preventing the bends: time.

Divers ascending from the depths of the ocean come up in a series of short lifts, spending a certain amount of time at each reduced depth in order to allow the dissolved nitrogen to come out of their blood slowly. The number of steps and the time spent at each is determined from a set of divers' tables.

Pilots of high-performance high-altitude military aircraft spend thirty minutes before takeoff breathing 100 percent oxygen through their oxygen masks in order to flush most of the nitrogen out of their bodies and thus prevent the bends from possible loss of cabin pressure.

Space shuttle crew members planning to engage in extravehicular activities (EVAs) in space suits are required to spend at least three hours breathing pure oxygen. This flushes nearly all of the nitrogen from their bodies before they venture outside in space suits for an EVA. This precaution was taken to be absolutely certain of preventing the bends in the event that the space suit developed a leak and lost pressure. The space shuttle space suit operated at a pressure of 5 psi while the Orbiter cabin was 14.7 psi. Bends were possible. It wasn't possible to rush any astronaut who suffered from the bends back to Earth. Therefore, three hours of pre-breathing were required by NASA aerospace medical experts to ensure that about 99 percent of the nitrogen was flushed, thus eliminating the possibility of the bends.

Space workers and tourists may encounter the bends in space if the pressurized module or their space suits suffer rapid decompression. If that happens, however, the bends will be only a painful part of a far more lethal and immediate problem of getting back into pressure at once.

Carbon Monoxide Poisoning

Carbon monoxide is yet another danger involved with living in a pressurized space environment. Carbon monoxide on Earth usually comes from the incomplete combustion of organic carbon-based fuels or materials.

Carbon monoxide is an odorless, colorless gas that has a tremendous affinity for attaching to red blood cells, an affinity 200 times greater than that of oxygen. Carbon monoxide poisoning therefore is a form of hypoxia and anoxia because it blocks oxygen from the red blood cells, which normally carry oxygen to body tissues. The onset of carbon monoxide poisoning is the same and just insidious as hypoxia.

Other than the exhausts of internal combustion engines, the most common source of carbon monoxide on Earth is cigarette smoke that contains about 3 percent carbon monoxide. A pack-a-day smoker usually has between 4 and 8 percent of his blood saturated with carbon monoxide, which is roughly half of the lethal concentration.

In space, carbon monoxide may come from electrical equipment that gets too hot and begins to smolder or burn in an oxygen atmosphere, from improper food cooking techniques, or from any number of sources that haven't even been imagined yet.

Summary

To recapitulate the human requirements for pressure and atmospheric composition necessary for breathing and thus life:

1. Humans require gaseous oxygen at a pressure of about 3 psi.
2. Carbon dioxide is exhaled by humans as a product of cellular combustion.
3. A careful balance between oxygen and carbon dioxide in a breathable atmosphere must be maintained so the human body can maintain the proper chemical balance.
4. Nitrogen is a diluent gas in the normal Earth's atmosphere, but its presence can create severe problems with evolved gas (the bends) if the atmospheric pressure is suddenly decreased.
5. An excess of any type of atmospheric gas can be hazardous.
6. Too much or too little atmospheric pressure can be hazardous.

Life-support systems can handle these requirements if they're working properly. But the knowledge of *why* life-support systems are needed is vitally important if a malfunction occurs or if a person encounters any number of hazardous emergency situations in space.

FIG. 4-1: *Although the thermometer was originally invented by Galileo in 1592, no accurate measurement of temperature was possible until the early 1700s when Fahrenheit developed the mercury thermometer and R.A.F. de Réamur invented the alcohol thermometer.*

CHAPTER FOUR

KEEPING COOL

Maintaining the proper atmosphere in space is only one of the require-ments to survival there. As on Earth, it's possible to be surrounded by an atmosphere with pressure and composition that's perfect for human needs, yet the temperature of that atmosphere can be too hot or too cold to live in.

Temperature control in space, however, is more than just heat-energy man-agement. As on Earth, the matter of humidity is also important.

Earth's Thermal Environment

The Earth is unique among the planets of the solar system for many reasons, due to the composition of the atmosphere; its gravitational field, which pro-vides atmospheric pressure; and the temperature on its surface. This combi-nation of conditions allows the Earth's atmosphere to hold water in a vapor state.

Under normal Earth conditions, water vapor is invisible. However, when the atmosphere becomes saturated with water vapor and can hold no more, any additional water vapor suspended in the air condenses into droplets. When this happens, the water droplets can be seen as a whitish haze or as clouds.

Earth isn't the only planet to possess clouds of water vapor. But no other planet has such an abundance of water. Almost 70 percent of its surface is always covered with cloud layers.

Because humans have evolved on Earth with its water-laden atmosphere, the effects of temperature cannot be discussed without also considering rela-tive humidity.

Relative Humidity

"Relative humidity" is a term often used with little real knowledge of what it means and how it relates to human comfort.

Any volume or "parcel" of air is like a transparent sponge. It will hold only a certain amount of water vapor. When it won't hold any more, it is said to be saturated. The total amount of water vapor that a parcel of air can hold depends upon its temperature and pressure and is termed its "absolute humidity." This is expressed in terms of weight of water per volume of air—i.e., grams per cubic centimeter or ounces per cubic foot.

For many reasons, a parcel of air is rarely completely saturated with water vapor. The actual amount of water vapor in it at any given time can be expressed conveniently in terms of the percentage of the absolute humidity. This is known as relative humidity.

The warmer a parcel of air becomes and the greater its pressure, the more water vapor it can hold. If the air parcel is cooled, its water vapor will condense to form fog, haze, clouds, and, in extremes, water that condenses out into large water droplets that fall as rain, freezing rain, sleet, or water crystals called snow.

Hot air will hold more water vapor than cold air because water vaporizes more easily at higher temperatures. Thus, an air parcel at 90 degrees F (32.2 degrees C) holds more water vapor than a parcel of equal volume and pressure at 30 degrees F (−1 degree C).

When an air parcel is cooled to the temperature at which it becomes saturated (also known as the "dew point"), the excess water vapor condenses in the form of fog, clouds, or precipitation. This can be demonstrated easily: on any hot summer day, note how the cold surface of a glass of ice water collects condensation on its surface. The cold surface causes the air next to it to cool, lowering its temperature to the saturation point in terms of the amount of water vapor in the surrounding air. The water vapor condenses on the glass surface.

As warm, humid air rises from the Earth's surface, it cools and the water vapor in it condenses.

The process is reversible. When water goes from the liquid state to the vapor state, it absorbs heat and cools any object with which it is in contact. When water evaporates into the air, the air is cooled along with the surface from which the water evaporated. Although water can't evaporate into air that is already saturated, the drier a parcel of air is—i.e., the lower its relative humidity—the easier it is for water to evaporate into it.

This is the principle behind the air-conditioning unit known as the "swamp cooler" that is used in the American Southwest. Hot, dry desert air is drawn through fibrous pads wetted with water that evaporates into the dry air, lowering its temperature by as much as 30 degrees F but also increasing its relative humidity.

Because of the cooling effect of evaporation, relative humidity is extremely important for human comfort and strongly affects human temperature tolerance.

Measurement of Relative Humidity

An instrument called a "psychrometer" is used to determine relative humidity. This device has one ordinary thermometer alongside an identical thermometer whose temperature bulb is covered with a damp cloth sleeve. The dry-bulb thermometer measures like any other thermometer while the damp bulb thermometer shows the temperature produced by the evaporation of the water in the cloth sleeve.

Relative humidity is then determined by reading the wet-bulb and dry-bulb temperatures and consulting a complex psychometric chart such as that shown in Figure 4-2.

Body Temperature Control

The human body is extremely sensitive to heat and works best between very narrow limits of temperature. It tries to maintain an internal temperature of 98.6 degrees F (37 degrees C), especially in certain highly specialized and critical portions—such as the brain—that *must* remain at a steady temperature if they are to work right. Some parts of the human body can get colder or hotter than 98.6 degrees F (37 degrees C) without permanent damage, but if the limbs and extremities get too cold they can become frozen.

Like a home heating system, the human body has a built-in temperature sensing and control system. When the body gets too hot, temperature-sensing nerves automatically open the pores of the skin to release perspiration, which then evaporates to cool the body.

If the body gets too cold, the pores close and the skin puckers up into goose pimples or "duck bumps." This is an instinctual reaction to cold that's inherited from the time when humanity's remote ancestors were covered with a thick coat of fur. They fluffed up their hair to provide additional insulation against loss of body heat by trapping air between the hairs of the fur. Today, the human body has no coat of fur that can be fluffed up to provide a thick mat of insulation. However, it still thinks the fur is there and tries to fluff it up. This causes goose pimples.

However, these are overt and observable human reactions to temperature. Other physiological changes occur because of temperature variations. Biologically, two temperatures are critical to humans: 42 degrees F (5.6 degrees C) and 89 degrees F (31 degrees C). This is because the human body loses most of the heat of metabolism not by sweating but by radiating it as infrared. Twenty percent of this comes from the arms, hands, and fingers. Regardless of the relative humidity, humans maintain their body temperature by radiation between 42 degrees F and 89 degrees F. The circulatory system is the main system that transfers heat from the body to the skin where it can be radiated into the surrounding environment. Blood returns from the fingers, hands, and arms through veins wrapped around arteries. This venous blood not only radiates heat through the skin but cools the outgoing arterial blood.

However, when the ambient temperature rises above 89 degrees F, the body can't radiate enough heat fast enough. Therefore, the venous blood returns to the heart through an emergency system, a vein network closer to the surface of the skin that also supplies water to sweat glands.

Below 42 degrees F, the venous blood also shifts to the emergency system to warm the fingers, hands, and arms.

The Temperature-Humidity Index

Because of this built-in human cooling system, one must know the temperature *and* the relative humidity when dealing with the effects of temperature. This has led to the use of an empirical number call the "discomfort index" or the Temperature-Humidity Index (THI) that is reported in the news media with the daily weather during the summertime.

THI is calculated by adding the wet-bulb and dry-bulb Fahrenheit temperatures together, multiplying the result by 0.4, and adding 15.

There is no rationale behind these numbers. They are "empirical"—i.e., they've been developed by rule of thumb from experience.

A THI of 75 means that most people are uncomfortable because of the summertime temperature. A THI of 80 or more means that almost everyone is uncomfortable.

Physical Effects of High Temperatures

However, above certain temperatures relative humidity plays only a minor role, if any, in human comfort.

At a temperature of 95 degrees F (35 degrees C), a person is uncomfortable at any humidity. When the temperature is over 102 degrees F (38.9 degrees

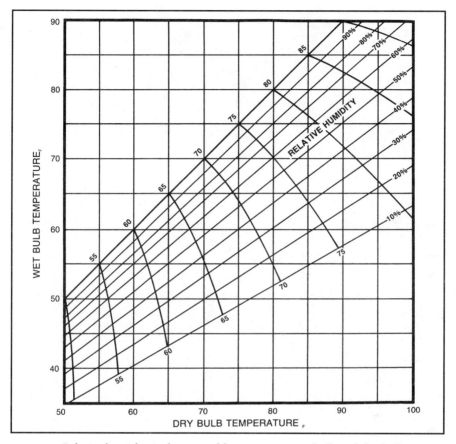

FIG. 4-2: *Relative humidity is determined by measuring wet-bulb and dry-bulb temperatures and referring to this "psychometric" or temperature-humidity chart.* (Art by Sternbach)

C), a person is acutely uncomfortable if no additional steps are taken to control body temperature.

In a place like Phoenix, Arizona, summertime temperatures often soar as high as 120 degrees F (48.8 degrees C). This is just *hot* and is bearable for short periods only because the relative humidity in such a desert climate ranges between 3 percent and 10 percent. Therefore, sweating helps cool the body. However, in New York City or Houston, a temperature of only 85 degrees F (29.4 degrees C) can be extremely uncomfortable when the relative humidity is between 80 and 90 percent, which is usual and normal. A person becomes soaked in perspiration that, because the air is already nearly saturated, can't evaporate fast enough to cool the body. It isn't the ambient temperature that affects a person as much as it is the inability to get rid of body heat.

Physical Effects of Low Temperatures

Many of the same effects are present in cold weather but are the results of somewhat different physical mechanisms.

In dry desert climates, winter cold seems milder although the temperature often drops well below freezing. This is because of the low humidity. When the air is dry, it cannot hold as much heat energy as when it's wet with water vapor. Therefore, cold, dry air cannot absorb as much heat from a human body.

In a cold, damp climate, winter weather can seem worse because the damp air can carry away more body heat faster.

Temperature and Heat

Temperature is not the same as heat.

Temperature should be thought of as the *speed* of molecules. The faster they move, the higher the temperature. Temperature is a measure of the degree of hotness.

Heat, on the other hand, is a general term meaning the quantity of thermal energy in a given amount of matter.

For example, a pound of water may contain the same amount of thermal energy as ten pounds of water but its temperature will be higher because the same amount of thermal energy exists in *less* volume of the same material.

Heat Transfer Methods

Heat is transferred from one parcel of matter to another by one or a combination of three methods: conduction, convection, and radiation.

Conduction heating is transferral of thermal energy by direct contact. A pan on a stove gets warm because it's in contact with the hot surface of the stove. Heat flows from the hot stove to the cooler pan.

Convection heating is a form of conduction heating. If water is placed in a pan on a stove, the water at the bottom of the pan is heated by conduction from the stove to the pan to the water. Being a fluid, the water becomes less dense as it's heated and therefore rises from the bottom of the pan because it's lighter than the cooler water surrounding it. This allows cooler surrounding water to move in to be heated by conduction. The heated water, while rising, passes some of its heat to water above it by conduction. Convection heating is thus conduction heating speeded up by a mixing process.

Radiation heating is something totally different. A person sitting in front of a heat lamp or basking in the summer sun on a beach isn't being warmed by conduction or convection. The lamp, the sun, or any warm body gives off heat in the form of infrared radiation. This radiation is the same as radio waves, microwaves, light, X rays, or other electromagnetic radiation except for its frequency. Infrared requires no medium for transmission. At the surface of a hot body, heat energy becomes infrared radiation which travels away from the body at the speed of light. When this infrared radiation reaches another body, it's converted back into heat energy.

A person can be warmed or cooled by a combination of these three methods.

On a cold day, a person who sits on a radiator to warm up gains heat by conduction.

When sitting in a tub of hot water, a person is heated or cooled by convection, depending upon the water temperature.

An electrical heater or the sun warms a person by radiation.

Air Movement Effects

The motion of air over a human body has a great effect on comfort because of the principles of conduction and convection heating. Air motion provides forced convection.

On a hot, humid day, a breeze or the air movement created by a fan removes the saturated air next to the body, replacing it with fresh air that hasn't been saturated by water vapor from perspiration.

On the other hand, a combination of low temperature and air movement will make a person feel colder than the actual temperature for the same reason. The wind strips away the layer of air next to the body even if heavy clothing is worn. This creates a wintertime problem known as "wind chill."

A table of wind chill (Figure 4-3) has been developed empirically in the same manner as the THI. It indicates, for example, that the temperature of 20 degrees F with a wind of 20 miles per hour will cause body heat loss equivalent to that in a −10 degree F temperature with no wind. In other words, the wind makes 20 degrees F feel like −10 degrees F.

Human Thermal Tolerances

THI, wind chill, and other factors relating to the high and low limits of human temperature tolerance weren't known until the Department of Defense conducted research. A soldier in the arctic or a flier in the stratosphere must be clothed properly or he will sit around trying to keep himself warm instead

Cooling Power of Wind on Exposed Flesh Expressed as an Equivalent Temperature												
Estimated wind speed (in mph)	Actual Thermometer Reading (F.)											
	50	40	30	20	10	0	−10	−20	−30	−40	−50	−60
	EQUIVALENT TEMPERATURE (F.)											
calm	50	40	30	20	10	0	−10	−20	−30	−40	−50	−60
5	48	37	27	16	6	−5	−15	−26	−36	−47	−57	−68
10	40	28	16	4	−9	−24	−33	−46	−58	−70	−83	−95
15	36	22	9	−5	−18	−32	−45	−58	−72	−85	−99	−112
20	32	18	4	−10	−25	−39	−53	−67	−82	−96	−110	−124
25	30	16	0	−15	−29	−44	−59	−74	−88	−104	−118	−133
30	28	13	−2	−18	−33	−48	−63	−79	−94	−109	−125	−140
35	27	11	−4	−19	−35	−51	−67	−82	−98	−113	−129	−145
40	26	10	−6	−21	−37	−53	−69	−85	−100	−116	−132	−148
Wind speeds greater than 40 mph have little added effect.	LITTLE DANGER (for properly clothed person) Maximum danger of false sense of security.			INCREASING DANGER Danger from freezing of exposed flesh.			GREAT DANGER					

Source: NAVMED Bulletin 5052-29

FIG. 4-3: *Wind-chill table. The higher the air velocity, the lower the apparent temperature and the greater the "wind chill." The chart is valid not only for Earth but also inside spaceships and space facilities.* (Art by Sternbach)

of attending to his duties. A soldier can't be sent into the tropics or the desert in a fur coat and be expected to function. A human being can't perform at top efficiency if the body is too hot or too cold. Therefore, to learn about the temperature tolerance of human beings, the armed services made an enormous number of experiments concerning the effects of temperature, humidity, and ventilation using thousands of volunteers as experimental subjects. Data were taken with the volunteers fully clothed, optimally clothed for the temperature-humidity-wind environments, and unclothed.

The data from these experiments have been compiled for several decades. In analyzing the results, researchers gained an increasing knowledge of human tolerance and reaction to what they call "thermal stress."

Comfort is complex and subjective. It's not a simple subject to deal with, as shown by the "comfort chart" (Figure 4-4) that relates dry-bulb temperature and relative humidity. Thermal stress occurs outside the normal zones of comfort.

If a person gets too cold, the chemical processes in the cells slow down and the cells eventually die. If body liquids freeze, they crystallize and destroy the fine structures of the cells.

On the other hand, if people get too hot, they can literally burn themselves up. If the heat generated by organic combustion cannot be removed from the body, the complex organic molecules inside the cells break down and cease to function.

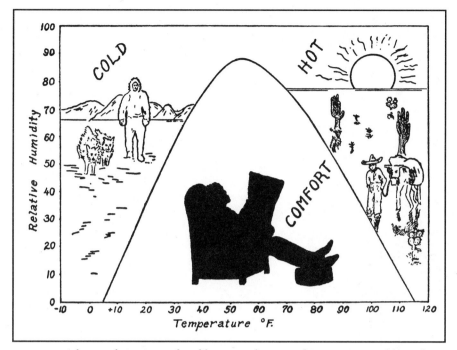

FIG. 4-4: *A human being is comfortable in a wide range of temperature and humidity conditions but uncomfortable outside those limits.*

The Human Heat Engine

The human body is a heat engine that combines the oxygen from the lungs and the food from the digestive tract, producing carbon dioxide, water, and heat.

Heat is measured in calories. A calorie is the amount of heat required to raise one gram of water one degree C (0.02 ounces of water one degree F). This is not the same calorie that is counted by people on a diet. A nutritional Calorie (capitalized) is equal to 1,000 thermal calories.

At rest, the human body gives off nearly 1,600 calories per minute. During active periods, the thermal output jumps to about 2,900 calories per minute. In comparison, the human body gives off about as much heat as a 100-watt light bulb. It's easy to understand, therefore, why a room full of people grows warm just from the accumulating body heat of everyone there.

The total heat created by the combustion of food is about 3,000 Calories per day, assuming a daily routine or schedule as shown in Figure 4-5.

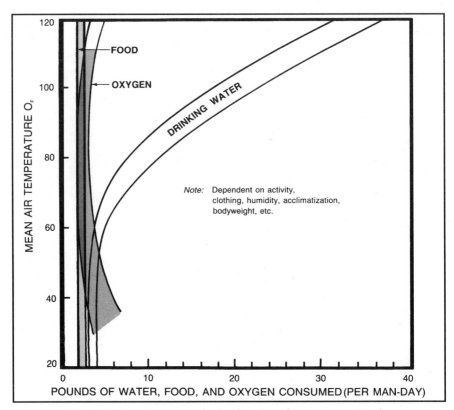

FIG. 4-5: *Human food and water needs as a function of temperature.* (Art by Sternbach)

When resting, about 78 percent of the heat is given off by normal heat transfer processes such as conduction, convection, and radiation. Twenty-two percent is given off by the evaporation of perspiration.

While working, 45 percent of the heat is lost by normal heat transfer methods and 44 percent through perspiration. The remaining 11 percent goes toward the actual accomplishment of physical work.

Calculations based on these numbers indicate that the human being is approximately 11 percent efficient or about as efficient as an internal combustion gasoline engine.

They also assume that the person is in an optimum temperature-humidity environment—not too hot and not too cold.

Because of these tests on human volunteers made for military purposes, human temperature and humidity tolerances as well as the absolute limits of temperature that a human can withstand are now well known and thoroughly documented. This is a classic example of swords being beaten into plowshares. The test results have had an enormous influence upon the design of clothing for outdoor and sporting purposes.

Clothing

Proper clothing is absolutely mandatory for human survival anywhere—on Earth or in space.

As discussed in Chapter Two, *Homo sapiens* has evolved as an unspecialized hunting species. Long ago, our ancestors gave up the insulating protection of a hairy coat. Humans retain only remnants at critical locations on their bodies where body parts must maintain proper temperature under the worst environmental circumstances. Hair acts in conjunction with sweat glands for cooling parts of the body. Hair serves to increase the evaporative surface for perspiration under conditions of high ambient temperature while serving as an insulating layer to trap air under conditions of low ambient temperature.

However, our ancestors learned how to provide themselves with artificial coats by using the hair and skins of other animals. This invention permitted them to quickly adapt to different ambient temperatures as well as to provide protection against solar radiation. Rather than be trapped by the environment, they learned how to take a comfortable environment with them wherever they went. As a result, they conquered climate and were able to spread across the entire Earth from the equator to the poles.

Humans are now doing the same thing in the void of the solar system.

Clothing is one of the great and continuing inventions of humanity. It has evolved locally into garb that will protect an Eskimo against the arctic climate as well as the Arab against the harsh desert environment. Clothing isn't just ethnic; it's pragmatic. It's been developed and perfected century by century by local people and is quickly adopted by other people who move there. The parka, the haik, the kaftan, the muumuu, the overcoat, the cowboy hat, the greatcoat, the kepi, the turban, the T-shirt, Bermuda shorts, boots, the wool sweater, gloves, earmuffs, sweatbands, pith helmets, and long johns are all incredibly sophisticated low-technology products brought about by centuries of testing and improvement. The oldest industry and the first to be touched in the industrial revolution was the clothing industry, along with its innovative testing area, the fashion industry.

Without clothing, the environmental range in which humans could live probably wouldn't extend far beyond the dry, temperate prairie-steppes where clothing is primarily worn for protection against chafing rather than as portable environmental control.

Human Temperature-Humidity Tolerances

Figure 4-6 shows the temperature tolerances of optimally clothed people as a function of temperature and time. This data holds true in space just as it does on Earth because it deals with *Homo sapiens* Mark One Mod One Version 1.0.

Properly clothed, a person can indefinitely withstand temperatures between −30 degrees F (−1 degree C) and 120 degrees F (48.9 degrees C). Note that the high end of the temperature tolerance depends upon the relative humidity.

Temperature tolerance, as indicated by Figure 4-6, is time-dependent. For short periods of time and if the person is acclimatized or used to it, a wider range of temperatures can be tolerated—from about −80 degrees F (−62.2 degrees C) to more than 140 degrees F (60 degrees C).

A temperature of 160 degrees F (71.1 degrees C) is tolerable for about 30 minutes, but a person would be capable of only 10 minutes of activity at 150 degrees F (65.5 degrees C) and about 5 minutes of activity at 200 degrees F (93.3 degrees C).

A temperature of 120 degrees F (48.9 degrees C) is tolerable for about an hour but is considered to be above the range of continual physical or mental activity.

Actually, mental activities begin to slow down, errors begin to creep in, and complex performance begins to deteriorate at temperatures above 85 degrees F (29.4 degrees C) regardless of the humidity.

Physical labor becomes extremely fatiguing at temperatures above 75 degrees F (23.9 degrees C).

On the low end of the temperature scale, physical stiffness of the arms and legs begins at 50 degrees F (10 degrees C). Cold injury to extremities is a function of time as shown in Figure 4-6.

Although some temperature-humidity tolerances are shown in Figure 4-6, a low humidity of 15 percent or less at any temperature can cause drying of external body fluids and the membranes of the nose and mouth. On the high end, most people generally consider humidities in excess of 90 percent to be intolerable.

The most comfortable temperature-humidity region for unclothed human beings lies between 65 degrees F (18.3 degrees C) and 90 degrees F (32.2 degrees C) with relative humidity between 30 percent and 40 percent and air movement velocity between 30 and 50 feet per minute.

Beyond these limits lie the hazards of hyperthermia (temperature too high) and hypothermia (temperature too low).

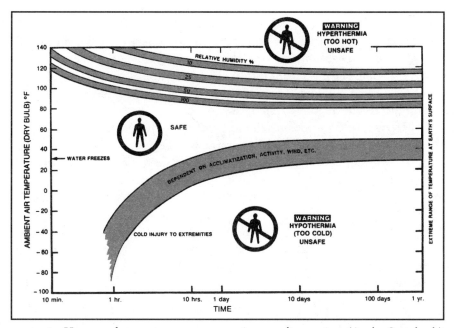

FIG. 4-6: *Human tolerance to temperature varies according to time.* (Art by Sternbach)

Effects of High Temperatures

Hyperthermia includes a number of conditions all caused by the human body becoming too hot.

One of these conditions is heat prostration, caused by failure of the dilation of the blood vessels of the skin. This prevents blood from acting as a coolant to carry away excess heat, which is then exchanged with the air in the lungs and exhaled if it is not radiated from the body from the skin. The symptoms of heat prostration include weakness, dizziness, vertigo, headache, nausea, blurred vision, and mild muscular cramping. The skin becomes strangely cold and the person begins to sweat profusely although the body temperature remains normal. This is usually a transient condition and can be easily treated by having the person recline in a cool environment, loosening or removing clothing, and drinking copious amounts of cool water to replace sweating-caused fluid loss. Rarely does heat prostration progress to the point of circulatory collapse, which would require the immediate attention of trained medical personnel.

Heat cramp has long been experienced by workers. It's called Stoker's Cramp, Fireman's Cramp, Miner's Cramp, and Cane-Cutter's Cramp. It's almost certain there will be an Astronaut Cramp, too. Muscular cramps are the result of failure of the body to replace sodium chloride and other minerals lost through the sort of profuse sweating that accompanies heavy physical work. Heat cramp happens fast and is painful. But it is transient and easily treated by relaxing in a cool place, drinking lots of water, and replacing minerals lost through perspiration to regain the body's chemical balance.

Hyperexia is perhaps the most hazardous form of hyperthermia because it's a profound disturbance of the body's basic heat-regulating mechanism. Initially, the symptoms of hyperexia or heat stroke may resemble those of heat prostration combined with heat cramp *except* the heart rate rises to 160 or more, there is no sweating, body temperature rises rapidly to 105 degrees F (40.5 degrees C), and the skin becomes hot, flushed, and dry. Hyperexia is an *emergency situation* that must be treated immediately to prevent profound shock, convulsions, cardiac failure, brain damage, and death. Hyperexia is a serious threat to life. Mortality may run as high as 20 percent. Heroic measures must be instituted immediately. The initial task is to reduce body temperature by immersion in an ice-water bath or by wrapping the body in a blanket soaked in cold water. The services of a doctor are mandatory because of possible changes in blood electrolyte balance. Hypothermia may occur once the hyperexia condition is reversed and therefore the constant attendance of trained medical personnel is required.

Effects of Low Temperatures

Exposure to cold conditions below the tolerance level may result in one or a combination of hypothermic injuries. The extremities are primarily affected; and injuries may involve function, structural disturbances, or failure of small blood vessels or nerves. Body tissue itself may become frozen, causing crystallization of the water in the affected cells. Hypothermia (sometimes called "exposure") may be accidental or acute, which leads to a depression of body temperature and a failure of the body's temperature-regulating mechanisms.

Exposure to high-humidity cold may result in frostnip (mild frostbite), chilblains, or trench foot while exposure to dry cold usually causes frostbite and acute hypothermia.

Treatment of hypothermic conditions involves slowly rewarming the body with warm baths, hot water bottles, and heating pads. Such rewarming must be done with water temperatures no higher than 110 degrees F (43.3 degrees C) to prevent burns or scalds. Tissue damaged by frostnip or frostbite should be dressed and a tetanus booster administered. Full treatment must be handled by a doctor.

This is particularly true of acute hypothermia which, like hyperthermia, is an *emergency condition* requiring immediate treatment, in this case, by warming. During treatment and recovery, attendance by a doctor or medical personnel is required because, in addition to possible tissue damage caused by freezing, there may be problems with metabolism and electrolyte balance.

Control of the Thermal Environment

Like the environmental factors of atmospheric pressure and composition, control of the temperature and humidity is an environmental requirement for space living. In essence, people must take their familiar terrestrial environment along with them. If they don't, they face the possibility of discomfort, loss of efficiency, and the hazards of hyperthermia and hypothermia.

The environment must have a temperature between 65 degrees F (18.3 degrees C) and 90 degrees F (32.2 degrees C) with a relative humidity range between 30 percent and 50 percent. Outside this limited temperature-humidity range, human performance begins to deteriorate and, if the temperature departs too far from the tolerable range, serious physiological problems can result which may, in the extreme, lead to death.

Careful design of the proper life-support system of the spaceship or habitat will create and maintain this temperature-humidity environment. Many technical solutions exist to accomplish this. The actual system used will depend upon the type and size of the spaceship or habitat and where it is located. No particularly difficult technical problems are involved in the design, construction, and operation of such systems. People started using them when the first cave fire was lit to warm this primitive living area.

Maintaining comfortable temperature-humidity environments is nothing new. People have been living in extremely harsh terrestrial temperature-humidity environments for millennia. Snug in their terrestrial homes and offices, people are no longer at the mercy of their environment. In space, however, no one can take these commonplace environmental factors for granted.

Centuries of accumulated know-how and more than enough scientific data and technical knowledge exist to build such temperature-humidity life-support systems. In addition, people have been living and traveling in space since 1961, and the technology of space life-support systems has been evolving ever since.

FIG. 5-1: *NASA Astronaut Scott Carpenter under acceleration in a centrifuge.* (NASA photo)

CHAPTER FIVE

ACCELERATION

Because of the development of transportation devices with great speed capabilities in the last 200 years, human beings have been subjected to increased levels of a force that has always been around: *acceleration*. People who live in space must be thoroughly familiar with acceleration forces and their effects on the human body because space transportation devices are the fastest machines ever developed.

However, it isn't *speed* that creates the various physiological effects of acceleration. There is a considerable amount of basic truth in the quip, "It isn't speed that kills; it's the sudden stop."

When the first modern transportation system—the steam-powered railway—was put into use in the 1830s, many people believed that passengers couldn't withstand the "horrendous speed of 30 miles per hour" attained by the early railway trains. It turned out that human beings can withstand any speed if they're in an enclosed compartment protected from the blast of air rushing past.

Compared to the stagecoach and canal barge, the railway train was able to change speed rapidly. This was the aspect of faster travel that had a demonstrable effect on people. Jolting starts and sudden stops—to say nothing of the sudden change in speed that's a part of any collision—tossed passengers around with forces much higher than those created by being thrown from a horse.

Human beings have no sense organ or physiological mechanism that can detect speed. But people do sense and react to a change of speed or *acceleration*.

What Is Acceleration?

Acceleration is defined as a change of velocity per unit time.

Although many people may think of speed and velocity as synonymous, they are not. Velocity is a function of both speed *and* direction. Velocity can be changed by altering either speed or direction.

When an object changes velocity, it accelerates. Everything inside of the object or attached to the object and moving with it is subjected to the same acceleration. This can be sensed as a force basically caused by the inertial property of mass—i.e., anything that's in motion wants to continue moving at that speed in a straight line. Therefore, a change of speed or direction produces a force opposing the change.

Sensing Acceleration

If an automobile is made to increase speed suddenly when the driver steps on the accelerator pedal, the occupants of the car are pushed back into the seat cushions. If the brakes are applied to slow down the car quickly, they'll be thrown forward. When a car is made to turn quickly, its direction has been changed and this is felt as a sideward force. These are mild forms of acceleration that don't greatly affect the human body. They are well within human acceleration tolerances.

But when velocity is changed rapidly, as in a car crash, passengers are subjected to high acceleration forces in a short period of time. These are also the sorts of accelerations that can be experienced in a spaceship.

A spaceship is just like an automobile in this respect. It's capable of accelerating. It may be capable of traveling at a speed of several miles per second, but it takes time to reach that speed. The amount of time and the change of velocity that occurs during that period of time are the factors that define acceleration force magnitudes.

Passengers in a spaceship that is changing speed feel the acceleration force just as they do in an automobile. But the acceleration forces involved in space travel can be five to ten times as high as those of an automobile and are of the same magnitude as the acceleration forces encountered by pilots and passengers in airplanes engaged in aerobatics.

Anyone traveling in or living in space must have a basic understanding of acceleration. On Earth, acceleration is a "take-it-or-leave-it" situation because it doesn't seem to be an important factor in everyday living. In fact, living on the surface of Planet Earth gives people a distorted mental image of the way the rest of the Universe works. People get used to the unique characteristics of planetary living that includes the constant presence of a gravitational pull plus the ability to keep acceleration forces within tolerable ranges of magnitude and duration—unless they fall or have an automobile accident.

Acceleration and Gravity

The effects of gravity are similar to those of acceleration. In fact, without special measuring instruments, a person cannot tell the difference between acceleration and gravitation. This is the "principle of equivalence" discovered by Albert Einstein. It's easily understood once explained.

If a person is in an enclosed elevator with no way to receive information about what's going on outside, it wouldn't be possible to tell whether gravity or acceleration was holding that person's feet to the elevator floor or if the elevator was accelerating upward with an acceleration that produced a force equal to that of gravity.

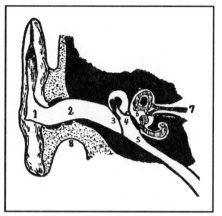

FIG. 5-2: *The semicircular canals of the human ear will detect acceleration or rotational motion. Key: (1) ear or* pinna, *(2) external auditory canal, (3) eardrum, (4) middle ear containing the ossicle bones, (5) Eustachian tube to throat, (6) inner ear and semicircular canals, (7) auditory nerve to brain, and (8) parotid salivary gland.* (Federal Aviation Administration)

A gravitational field accelerates mass. No one really knows *why* a gravitational field does this. No one really knows what gravity is, although there are many scientific hypotheses and theories about it. The gravitational field of Planet Earth always pulls objects toward its center. If the ground didn't get in the way, an object would fall to the center of the Earth. The ground surface prevents this, however, by stopping the fall and thereby exerting a force on the object that cannot be sensed as any different from acceleration force.

Measuring Gravity

For convenience, engineers use the term "g" or "gee" for measurement of acceleration because this compares the forces to the pull of Earth's gravity on the planet's surface.

Standing on the Earth at sea level, everyone experiences an acceleration force of 1 g. This is equivalent to being accelerated with a velocity change of 32.17405 feet per second every second. This is usually rounded off and written as 32.17 ft/sec/sec or 32.17 ft/sec^2. This gravitational acceleration isn't the same everywhere on Earth and decreases as altitude above the Earth's surface increases. It's only when space pilots have to compute highly accurate spaceship flight paths near the Earth that the actual gravitational acceleration numbers must be used in the calculations. Therefore, although the gravity field of the Earth varies, the Earth's "standard gravitation acceleration" of 32.17 ft/sec^2 is used for most rough "back of the envelope" calculations.

Two g's would be an acceleration equal to twice that of gravity or an acceleration of 63.34 ft/sec^2.

This can be illustrated if a person gets on the bathroom scale and the scale shows 150 pounds. It means the person is being pulled toward the center of the Earth, exerting a force of 150 pounds on the scale because of gravity. If Earth's gravity were twice as strong—i.e., if it were 2 g's—the bathroom scale would show the person's weight as 300 pounds.

Now take the scale and get aboard an elevator with the person. At rest in the Earth's gravity field, the scale would still show the person's weight as 150 pounds. If the elevator is started upward and gains speed at 32.17 feet per second every second, the person would be subjected to 2 g's of acceleration—1 g from the Earth's gravity field and 1 g from the acceleration of the elevator. The person's body would feel as if it were subjected to 2 g's and weighed 300 pounds.

Acceleration Effects

Automobiles rarely achieve accelerations of 1 g, but airplanes can produce accelerations of 10 g's or more while turning. Rocket vehicles can generate accelerations as high as 100 g's.

Electronic instruments and other mechanical devices can be built to withstand the crushing force of 100 g's, which makes them seem to weigh a hundred times normal, but people can't.

At 100 g's acceleration, a person would weigh 100 times normal. Bones and muscles cannot withstand these forces, and a human being under that sort of acceleration is crushed by the increased apparent body weight.

The Increasing Need for Data

Most of the research on human tolerance to acceleration was carried out between 1945 and 1965. Military pilots during the Second World War re-

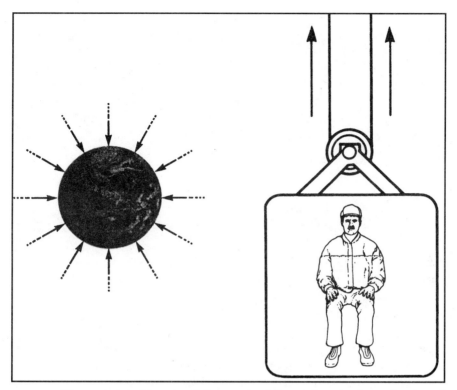

FIG. 5-3: *If a person is inside a closed elevator, the difference between the acceleration of gravity and the acceleration caused by change of velocity feels the same.* (Art by Sternbach)

ported that the accelerations attained during combat maneuvers were causing them to lose consciousness. Pilots discovered that if the acceleration force pulled them down in their seats, they'd "black out." The acceleration force pulled the blood from their heads, temporarily starving their brains of oxygen. This resulted in a loss of vision and, if the acceleration was great enough or continued too long, unconsciousness. Acceleration in the opposite direction created when performing an outside loop would cause blood to rush to their heads, making them "red out" instead. Accelerations that pushed them back into their seats ("eyeballs in" acceleration) was easier to withstand than the opposite ("eyeballs out"), because their seats supported their bodies during "eyeballs in."

But no one knew how many g's a pilot could withstand before his mental and physical abilities were impaired to the point at which he couldn't fly and fight. Military requirements to build faster and more maneuverable combat

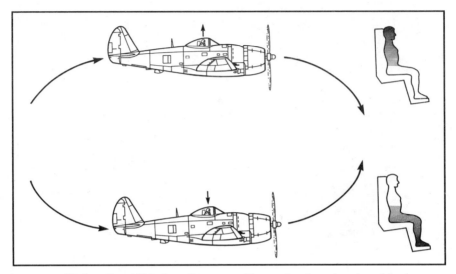

FIG. 5-4: *"Red-out" and "black-out" are caused by accelerations that draw blood away from the human brain.* (Art by Sternbach)

airplanes therefore required designers to have more data on acceleration tolerances to prevent the airplanes from killing their pilots.

Engineers who conducted early, pre-*Sputnik* (before 1957), studies of space travel and then designed the first manned space capsules needed information on how much acceleration a person could withstand. Because the early space boosters were converted military ballistic missiles, the g-forces on any of the space capsules and their occupants boosted by such rockets could be as much as 9 g's. Could a human being withstand such acceleration for periods of several minutes? Would the astronaut be able to exercise control of the space capsule during the high acceleration of launch? What was the best position for the astronaut with respect to the acceleration direction? How much acceleration could an astronaut withstand during reentry and landing? The answers to these questions determined many design features of the early Mercury and Gemini space capsules, down to the size of the parachutes required to keep landing accelerations and shocks to an acceptable level.

(The first US space capsules landed in the ocean to keep the landing forces within human tolerance. The Soviet *Vostok* space capsule landed so hard that the cosmonaut had to eject from the descending capsule and come down on a separate personal parachute.)

As body-contact sports such as football became more rugged and physi-

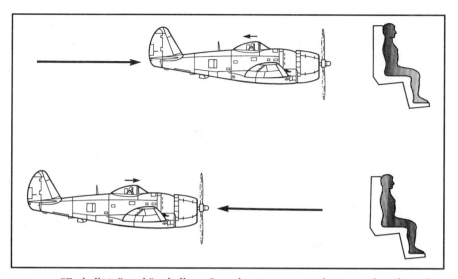

FIG. 5-5: *"Eyeballs-in" and "eyeballs-out" accelerations are graphic terms describing the effects of fore-and-aft accelerations on a human being.* (Art by Sternbach)

cally stressful, indications showed that impacts were injuring athletes in spite of protective padding. What were the human tolerances to short-duration accelerations with a very high rate of onset of acceleration ("jolt")? Data was obviously needed to completely redesign much existing athletic equipment and new equipment as well.

Finally, the increasing number of automobile accidents in which the impacts caused serious injuries and death led the federal government to demand safer cars that wouldn't kill their occupants by collapsing and crushing them or by throwing them violently against sharp interior fittings such as knobs, steering wheels, or decorations. Furthermore, the United States Air Force wanted to know how much g-force a pilot could take before losing control of the plane. The Air Force also discovered it was losing more pilots in automobile accidents than in airplane crashes.

These demands for information required a better understanding of human tolerance to acceleration and to "jolt" or a change of acceleration.

Getting Acceleration Data

Before astronauts and cosmonauts began riding rockets into space, research was carried out on Earth to determine the limits of human tolerance to acceleration. Researchers used centrifuges that spun volunteer human subjects

FIG. 5-6: *Huge whirling centrifuges are used to test human tolerances to long-term effects of moderately high accelerations.* (North American Aviation, Inc.)

around in capsules at the end of long mechanical arms, thus creating high and sustained accelerations by centrifugal force.

Centrifugal force is really an acceleration force caused by a constantly changing direction at a constant speed as shown in Figure 5-7. The person in the centrifuge cab has mass that possesses inertia which in turn wants to keep the mass going in a direction tangent to the circle of rotation. However, the centrifuge arm keeps pulling the cab and its human occupant inward, constantly changing the direction of the velocity.

Researchers also used linear horizontal accelerators. Some of these were rail-mounted sleds propelled horizontally by rockets along railwaylike tracks on the ground. They were stopped by scoops that extended into water troughs between the track rails. Other horizontal accelerators were moved rapidly by air-driven pistons to provide short durations of high accelerations whose data were useful in crash studies.

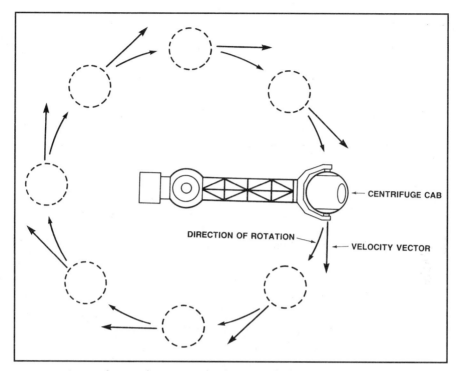

FIG. 5-7: *A centrifuge produces a true acceleration which is a change in the magnitude or direction of velocity by constantly changing the direction as shown here.* (Art by Sternbach)

Human Acceleration Tolerance Factors

It was generally agreed among experts in 1940 that 10 g's was the limit of human tolerance to sustained acceleration. No scientific or experimental evidence existed for this; it was an educated guess based upon reports of pilots who had been subjected to "high accelerations" of perhaps 6 to 8 g's in "pursuit" airplanes. During the Second World War, combat maneuvers actually created forces as high as 10 g's on pilots who reacted quite differently than the experts anticipated. The need for human acceleration research was compounded when high-speed jet airplanes first appeared.

Dr. John Paul Stapp, an Air Force flight surgeon, pioneered acceleration, jolt, and crash studies. He became famous because he wouldn't allow subordinates and volunteers to participate as experimental subjects unless he had personally gone first. Using centrifuges and rocket sleds, Stapp and other Air

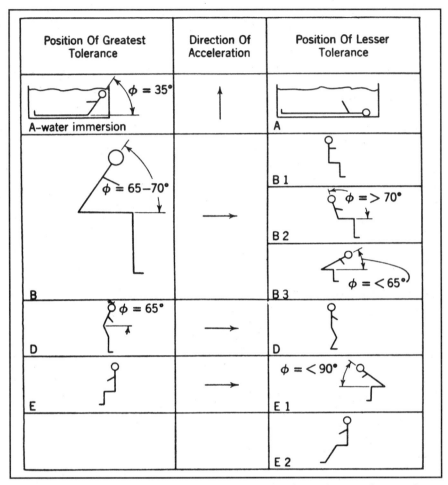

Position Of Greatest Tolerance	Direction Of Acceleration	Position Of Lesser Tolerance
$\phi = 35°$ A–water immersion	↑	A
$\phi = 65-70°$ B	→	B 1 $\phi => 70°$ B 2 $\phi = <65°$ B 3
$\phi = 65°$ D	→	D
E	→	$\phi = <90°$ E 1
		E 2

FIG. 5-8: *The position of a human being with respect to the acceleration direction has much to do with acceleration tolerance.* (US Air Force)

Force "human factors" researchers discovered that human tolerance to acceleration involves a combination of four main factors:

1. the *direction* of acceleration with respect to the human body
2. the *intensity* of the acceleration force
3. the *duration* of the acceleration force
4. the *rate of onset* or the suddenness with which the acceleration is applied to the human body.

These four factors are intimately interrelated.

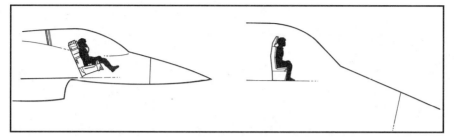

FIG. 5-9: *In both high performance fighter airplanes such as the F-16 (left) and the NASA space shuttle Orbiter (right), seats are designed and positioned so the accelerations can be easily tolerated by the human occupant.* (Art by Sternbach)

Constant Acceleration Tolerance Limits

Insofar as long-term human tolerance to acceleration is concerned, the primary limiting factors are the *direction* and the *intensity* of the force.

The *directions* of acceleration and their limits are shown in Figure 5-8.

A person has the greatest tolerance when lying down and immersed in water with the torso and head inclined upward at an angle of 35 degrees. A person lying flat actually has less acceleration tolerance.

Water immersion adds weight and complexity to any sort of travel in aircraft or spaceships. The best position for maximum tolerance is a reclining sitting position, upper legs and thighs at right angles to the acceleration direction, and the head and torso inclined at a slight angle against the direction of acceleration. This position has been universally used in crewed space vehicles to date. In the USAF Lockheed-Martin F-16 "Fighting Falcon" jet fighter plane, the pilot reclines at a greater angle with his legs in the position shown in Figure 5-9.

Figure 5-10 shows the tolerances for acceleration expressed in terms of "g-minutes" (one g-minute equals 1 g applied for one minute). Figure 5-10 was compiled for the use of aircraft and spaceship designers for crew and passenger seating in various orientations—backward or "eyeballs out," forward or "eyeballs in," foot-to-head, and with water immersion.

A person standing erect as on Earth can withstand 2 g's almost indefinitely. It's like carrying someone else around—difficult but bearable. But a person becomes fatigued unless conditioned by practice and regular exercises.

It's possible to remain standing at an acceleration of 3 g's applied in a head-to-foot direction, but walking and climbing become impossible; even crawling becomes difficult and done only with great effort. At 3 g's the only comfortable position is sitting.

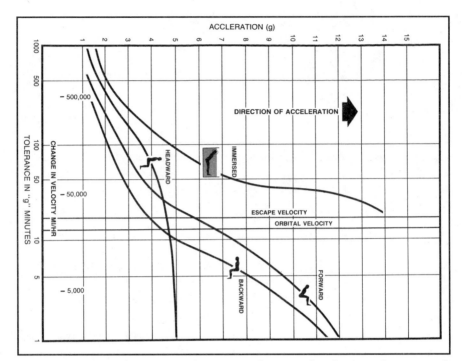

FIG. 5-10: *Human tolerance to acceleration as a function of acceleration, direction, and time.* (Art by Sternbach)

At 5 g's, only slight movements of the arms and head are possible. Visual symptoms such as "gray-out" begin to appear at 2.5 g's and blackout occurs at 7 g's. At 8.5 g's, the human heart can't force enough blood into the brain against acceleration, and unconsciousness results.

An acceleration of from 18 to 23 g's causes structural damage, especially to the spinal column.

With the acceleration in the opposite direction—foot-to-head or "head down"—the tolerance limits are *much* less.

However, if the acceleration is taken *across* the body rather than from head to foot, a constant acceleration can be tolerated for a much longer time at a high level. Human subjects have withstood constant accelerations of up to 17 g's in the supine position with no loss of vision or unconsciousness. Auditory function remains intact at 14 g's. Up to 30 g's can be tolerated in the supine position without structural damage.

Although head and arm movements become impossible at 6 g's, the wrists, hands, and fingers can move at accelerations up to 12 g's.

With the possible exception of military spaceships, it's unlikely that humans will encounter sustained accelerations of more than 3 g's in spaceships because their acceleration will be specifically limited to this level to provide for the comfort of human occupants.

In space travel, constant relatively long-term acceleration occurs over periods ranging from several minutes to an hour. As the duration of acceleration increases, human tolerance decreases. It might be possible for a human being to withstand 4 g's for several days, but it would be difficult to carry out normal biological activities. No information presently exists on the potential psychological effects to long-term exposure to high levels of acceleration.

On the other hand, as the duration of acceleration grows shorter, human tolerance becomes greater but another factor enters the picture: the rate of change of acceleration or "jolt."

If one second is required to go from 1 g to 10 g's acceleration, the rate of onset is 9 g's per second. This is a mild jolt. Applying 100 g's at the rate of 1,500 g's per second for a hundredth of a second is equivalent to the force of a fist banging on a table. However, 100 g's applied at 1,500 g's per second to the entire human body requires that the person be properly supported or protected against this sort of a jolt. As the rate of onset approaches 3,000 g's per second, extreme crash conditions are involved and physical damage such as the rupture of large blood vessels can become lethal.

Aerospace medical researchers have learned that high impacts and rates of onset are encountered by people in ordinary athletics. The author helped instrument two healthy college football players with devices to measure the acceleration and rate of onset they experienced when blocking and tackling one another. They experienced 70 g accelerations at a rate of onset of 2,000 g's per second for as long as 0.2 seconds. This data, taken in 1960, helped establish the Air Force criteria for ejection seats and escape capsules used in military aircraft.

Other data on human tolerance to acceleration and jolt have come from strange places. To get information on the strength of the human neck and to determine whether or not head restraints would be required in aircraft ejection seats, aeromedical researchers studied old books on hanging. They then wired an anthropomorphic dummy with instruments and conducted hangings from a gallows for the benefit of recording instruments. The weak-hearted members of the research team declined to carry the investigation further when they dropped the dummy ten feet and pulled its head off. At that point, however, they had learned the limiting strength of the human neck.

The absolute top of the acceleration tolerance spectrum is an oddity. A paratrooper whose parachute had failed landed on his back in the soft dirt of a recently plowed field and walked away with only a sprained wrist. He'd survived an estimated 200 g's for 0.015 seconds.

Protection Against Acceleration and Jolt

As a result of these tests, it was learned that a human being subjected to acceleration and jolt acts like a floppy bag of gelatin that gives under acceleration forces and bounces. Protecting a human against acceleration and jolt involves proper cushioning and support to distribute the forces over as much of the body as possible.

The human body must be kept from bouncing around inside the vehicle. Seat belts in automobiles and airplanes *do* make it much more likely that a person will survive an impact. When the cross-chest strap is added, the chances of crash survival improve even more. The best restraint harness has been used for decades in military and aerobatic airplanes: a lap belt, a strap going over each shoulder, and a crotch strap to keep the person from sliding out under the harness. This type of harness will restrain a human against very high acceleration forces in the eyeballs-out direction that's extremely critical in most crash situations.

Cabin and vehicle design has eliminated most of the sharp projections and edges that can cause injury if the body strikes them during the rapid acceleration and jolt of a crash. However, deeply padded surfaces can cause the body to rebound. In some cases, the best way to catch and cushion a human body is on a sheet of steel that will deform and stretch without rebounding, or into a plastic or honeycomb structure that will collapse and absorb the energy of impact.

It has continually amazed researchers to discover that human beings are more rugged than the devices they travel in. Studies of human acceleration and jolt tolerances have shown that, in nearly all cases, properly restrained and supported people have withstood crash forces only to be crushed by the vehicle structure collapsing on them. This has led to the re-design of automobiles as well as airplanes. And it's one area where aerospace research has paid off in everyone's lives.

Anyone who lives and works in space must be very conscious of acceleration at all times not only because of the accelerations experienced in space travel but because the space environment itself has a unique characteristic that can't be easily duplicated for more than a minute on Earth.

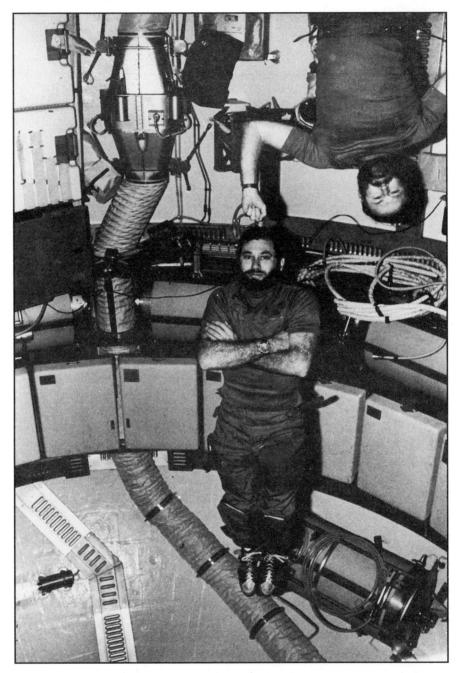

FIG. 6-1: *The NASA Skylab in 1973 gave people their first opportunity to move freely about in weightlessness.* (NASA photo)

CHAPTER SIX

WEIGHTLESSNESS

W EIGHTLESSNESS or "zero-g" is a characteristic of space that cannot be experienced by people on the surface of the Earth except for brief exposures of less than a minute.

Weightlessness is a consequence of how spaceships travel and also of the peculiarities of orbital mechanics. For most space travel and in most space facilities, the forces of gravity will appear to be absent.

The effects of weightlessness on the human body and even the mind are more profound than the effects of acceleration because people in space live and work in the weightless condition most of the time unless they're in a facility in which part of it rotates to produce the pseudogravity of centrifugal force.

The Reason for Weightlessness

Some people assume that weightlessness exists because the spaceship or station is beyond the pull of gravity or not in a gravitational field. Nothing could be further from the truth. Weightlessness is caused by a type of motion *within* a gravitational field.

Like the Earth's atmosphere, the Earth's gravitational field slowly decreases in strength or the ability to accelerate an object with increasing distance from the Earth until it finally becomes weaker than the gravity field of the Sun. In the Earth-Moon system, this means that nothing is ever beyond the influence of the Earth's gravity. In the solar system, nothing is ever beyond the pull of the Sun's gravity.

Why is a spaceship or facility in a weightless condition if it's never beyond the influence of a gravitational field?

The rocket-propelled spaceship that currently exists or whose design can be projected into the foreseeable future is an "impulse" driven vehicle. It achieves the necessary velocity to shape its path through space by the application of propulsive force that accelerates it during the first few minutes of flight. Once having achieved the proper velocity, its rocket propulsion system is then shut down. The spaceship then coasts through space toward its destination where a final application of rocket impulse is made to match velocities with the destination—planet, satellite, space station, or other spaceship.

This is the most *efficient* way to travel between any two points in space using the rocket engine as we know it today. It requires the least amount of rocket propellant and therefore energy.

The only difference between space travel in the Earth-Moon system and the solar system is the time involved to make the trip that in turn is a function of the velocity to which the spaceship can accelerate. But that's a subject for a book about rocket propulsion and celestial mechanics. Many books on those subjects have been published.

At some time in the future, other types of space propulsion systems may become available to permit "constant-boost" space flight—i.e., constant acceleration at, say, 0.1 g's out to the midpoint of the flight where the spaceship swaps ends and begins to decelerate at 0.1 g's to its destination. In constant-boost flight, there is no weightlessness or zero-g. And travel times become very short, even at accelerations/decelerations of a fraction of a g. A flight from the Earth to the Moon at a constant boost of 0.1 g's would require less than 11 hours and the maximum velocity at "turnover" would be 12.1 miles per second or 43,524 miles per hour. Constant-boost flight really cuts the solar system down to manageable size in terms of trip times. A flight to Mars under the worst possible planetary alignment conditions would require about 10 days under a constant boost of only 0.1 g's. At a constant 1 g boost, a Mars trip is only 4.5 days. However, one major problem exists with the propulsion system for a constant-boost spaceship: No one knows how to build it. Theories exist and a few gadgets may or may not be primitive "proof of principle" devices, but no one has yet demonstrated a "space drive." So rockets are the future of space travel for the foreseeable future.

Therefore, a spaceship that's coasting to its destination with its engines shut down or a space facility in orbit will fall freely in a gravity field.

In the case of a spaceship, it falls *up* away from the starting site. Approaching the Earth or another celestial body, it falls *down*. This is not unusual behavior in the Universe. Many objects fall both up and down.

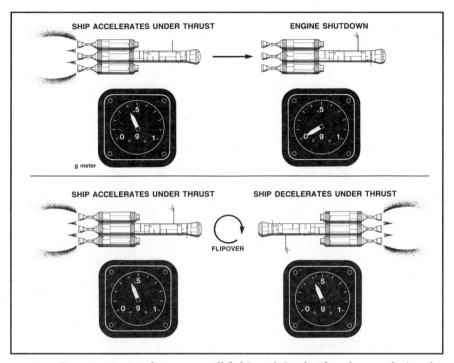

FIG. 6-2: *During space travel, a person will feel "weight" only when the spaceship's rocket motors are thrusting. In classical flights, acceleration occurs only at the start and end of the flight with weightlessness in between. Future spaceships will make "constant boost" flights with a flip-over at mid-point of the flight, thus creating pseudogravity acceleration during the entire trip.* (Art by Sternbach)

For example, when someone throws a tennis ball, rubber ball, handball, or baseball into the air, the motion of the throwing arm and hand imparts a velocity—speed and direction—to the ball after being accelerated over a distance of about 5 feet. The ball will, of course, travel to a much greater distance than the 5 feet through which it was accelerated.

The ball begins to fall the moment it leaves the thrower's hand because it's no longer being accelerated by the hand-arm motion. First the balls falls up, then it falls down. If the drag of the Earth's atmosphere is neglected, the only force acting on the ball in flight is the Earth's gravity. It draws the ball toward the center of the Earth.

Now imagine being inside the ball. Anything attached to the ball or inside the ball falls freely with it. There is no internal force that could be perceived as "gravity" acting between the ball and anything inside or on the ball.

The "throwing arm" of any spaceship or orbiting facility is the rocket-powered device that accelerated it to final velocity or to orbital velocity. Once a satellite is placed in orbit, it falls constantly toward the center of the Earth. But the Earth's surface curves away from it as fast as it falls. So it never reaches the Earth's surface. The fact that the Earth is spherical must be taken into account.

Newton's Cannonball Analogy

The best analogy for the concept of weightlessness was written by the person who originally worked out the mathematics of the law of gravity, Sir Isaac Newton. In his pioneering work about mechanics, *Philosophiae Naturalis Principia Mathematica* (known as the *Principia*), published in England in 1687—with the financial assistance of Sir Edmund Halley for whom the comet is named— Newton imagines a planet like Earth with a high mountain on its surface whose summit extends above the top of the planetary atmosphere.

If a cannon atop the mountain (Figure 6-3) fires a cannonball horizontally, the cannonball will begin to fall the instant it leaves the muzzle of the cannon. It will then strike the surface at some given distance from the cannon, depending upon the speed at which it was fired.

FIG. 6-3: *Sir Isaac Newton's explanation of satellite orbits using a cannon to launch shells at increasing velocity until one finally falls all the way around the world.* (Art by Sternbach)

If the cannon is reloaded with more gunpowder and the cannonball fired again with even greater muzzle velocity, the cannonball will go farther before it finally strikes the ground.

A condition exists, Newton pointed out, in which a large gunpowder charge in the cannon would give the cannonball enough muzzle velocity so it would go all the way around the planet and hit the cannon from behind. If the cannon is taken out of the way in time, the canonball will continue to go around the planet forever like a little moon.

This phenomenon is often explained by saying that the centrifugal force generated by the velocity of the cannonball exactly matches the gravitational force. This is also a true analogy. But the more general case of the constantly falling cannonball moving above the surface of a sphere that continually bends away from it provides a better explanation of the motion and forces of nearly all orbiting bodies.

The Sensation of Weightlessness

How does this produce the sensation of weightlessness?

When someone sits in a chair on Earth, the weight of the body against the chair can be felt. Actually, this is the feeling of the Earth's gravitational field attempting to accelerate the person toward the center of the Earth. If the chair, the floor, the building foundation, and the ground were not there, the person would indeed fall unrestricted to the center of the Earth. Because the acceleration of the Earth's gravity is 32.17 feet per second per second, the person would fall at a speed of 32.17 feet per second at the end of the first second, 64.34 feet per second at the end of the second, and so forth. The person would have the sensation of falling.

The weight that is felt while sitting in the chair is caused by the resisting force of the chair against the acceleration of the person's body by the gravity field. Remove that resistance and the person would fall freely in a weightless condition.

The Falling Elevator Analogy

To reinforce the explanation, imagine the person and the chair in the closed elevator car suggested in the previous chapter. If the elevator cables are cut and the elevator begins to fall freely down the shaft, the person would feel no weight force and would fall with the chair and the elevator, feeling no weight.

Earthly Experiences

Weightlessness is a condition encountered on the Earth only for short periods of time. People live with gravity day and night, and it's normal to them. They expect it to be there and really don't think about it. When it isn't there or when the sensation of a force between their bodies and the chair or floor disappears, they know they're falling. Alarm bells go off in their brains because they've learned from long experience that falls *always* end in sudden and sometimes painful stops. The longer the falling sensation continues, the worse the bump is going to be.

All life on this planet evolved with gravity and has adapted to it. Plants send their roots downward and their stalks upward in response to the tug of gravity, a phenomenon known as *geotropism*. Although sea animals gain most of their bodily support from buoyancy, land animals have skeletons and muscles to support their bodies against the force of gravity.

Occasionally, a person is subjected to a brief period of weightlessness when they fall freely—during the initial descent of a fast elevator, jumping off a diving board, bouncing on a trampoline, or skydiving with a parachute. However, most of these periods of weightlessness last only for a second or so.

In space, weightlessness is normal and the acceleration of gravity isn't.

The Effects of Weightlessness

When space flight began in 1961, not much was known about the effects of weightlessness on the human body, how someone would react to constant falling, or whether people could live without a gravity force acting on them. As the years have gone by and as more people have flown in and lived in space for increasing periods of time, more answers and data have become available. Some of the experiences have been positive and some haven't. But, in general, based on the studies and experiments made on people who have traveled and worked in space, here's what is known:

For short periods of time up to a year, people have no problem living in weightlessness. Some people, however, require several hours to several days to adapt to the weightless condition. The latest data show that a person has a 50 percent chance of becoming motion sick during the first day in weightlessness. They usually get over it after a few hours.

People have no trouble carrying out normal living functions in zero-g. Gravity isn't required for breathing. Food and water can be swallowed in weightlessness. Elimination of human wastes presents no serious problems if the "waste-management system" is designed to receive and contain such wastes.

Muscle coordination is somewhat affected by weightlessness at first. Hand-eye coordination suffers the most. During the Skylab 3 mission in mid-1973, American astronaut Owen Garriott reported that crew members found it almost impossible in the darkened sleeping quarters to reach out and touch the light switch located less than two feet away. "The result was not just a near miss," Garriott reported. "We found that our hands might first encounter a locker as much as forty-five degrees away from the proper direction. Although I tried to 'practice' this move on a number of occasions, I still could not do it well after two months."

FIG. 6-4: *Living and working in weightlessness can pose new problems as well as solving others. Astronaut John Young floats in the mid-deck of the NASA space shuttle Orbiter with personal items and food containers floating all around him. (NASA photo)*

However, if a person can see the surroundings, no trouble is encountered with hand-eye coordination in weightlessness. Both American and Soviet/Russian space crews worked successfully in weightlessness for months or more and reported no problems with hand-eye coordination if they could see what they were doing.

Furthermore, American astronauts had no trouble maintaining their orientation regarding up and down. According to former astronauts Gerald Carr and Ronald Evans, "up" is the direction their heads were pointing while "down" was toward their feet, regardless of their orientation.

American astronauts reported it was easy to move around inside a spaceship or facility in weightlessness. In fact, they found it was much easier to get around in weightlessness than it was in a simulator on the ground. Divers, gymnasts, or acrobats are quite capable of moving and controlling their bodies during brief periods of weightlessness. Pictures from American and Russian spaceships and facilities show that people adapt quickly to getting around and even invent weightless games and "astrobatics" (weightless acrobatics). It looks like a lot of fun, and astronauts reported it's one of the best recreations they've discovered in space. It doesn't take people long to overcome their instinctual falling reactions and start turning it into "fun," as a visit to any theme amusement park with wild rides will confirm.

Adapting to Weightlessness

From the very start of space flight, some people have adapted immediately to weightlessness while others have required several days. Fear of falling is a survival trait on Earth. The semicircular canals or otoliths in the middle ear don't know a person is in weightlessness and recognize it only as falling. The signals sent to the brain conflict with those from the eyes, causing confusion and producing a situation called an "autonomic storm."

Most people going into space haven't been able to overcome these fears and confusing sensory signals. As a result, they've suffered from motion sickness that includes nausea and disorientation. Usually, this weightlessness-adaptation syndrome disappears in a matter of minutes or, at the most, in a few days.

Space motion sickness was first reported by the fourth person in space, Soviet cosmonaut German Stepanovich Titov, who was launched in *Vostok 2* on August 6, 1961. Titov experienced an initial nausea and disorientation but was able to adapt to weightlessness within a few hours.

Some American astronauts haven't been able to totally adapt to the weightless condition outside the spaceship or facility during "extravehicular activity"

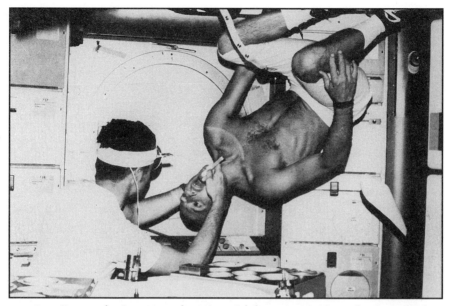

FIG. 6-5: *Because there's no up or down in weightlessness, activities such as a medical checkup become easier to do.* (NASA photo)

or EVA. Some of them suffered from a phobia allied to the fear of falling: acrophobia, or the fear of high places.

No one who has gone into space to date has been unable to adapt to weightlessness within a day, although some people have had a very difficult initial few hours.

Most people who have gone into space thus far have been very highly motivated. They worked hard, often for years, to get into space. Many of them were experienced test pilots. It was quite possible that their strong motivational drives enabled them to overcome and conquer their initial reactions to weightlessness. However, motivation may not be a factor. News reporters and other people who have not been extensively trained as astronauts and who have traveled into space seem to have overcome space sickness quickly, if they suffered from it at all.

Vertigo

Some disorientation may be due to vertigo. The word "disorient" means "to turn to the east" or, in the vernacular, "Which way is up?" Vertigo is a severe form of disorientation, a state of temporary spatial confusion resulting from the brain's misinterpretation of conflicting information received from other sense organs, primarily the eyes. When visual information doesn't match other information coming from the balance sensors in the middle ear—the otoliths or semicircular canals—and the kinesthetics of touch and muscle resistance, the brain doesn't know which to believe or how to sort out the conflicting information.

It's difficult to describe what vertigo is like because it amounts to almost total sensory confusion. It's possible to experience vertigo in amusement park rides on Earth. Airplane pilots operating under instrument flying conditions in clouds or at night can fall victim to vertigo.

Vertigo can be created on the ground by a rotating seat known as a Barany chair. It is used to acquaint airplane pilots with vertigo because its danger can be reduced if a person understands the nature and causes of the condition, avoids the environmental situations that cause it, and learns to heed and *believe* visual inputs from airplane instruments that monitor and report on the airplane's attitude and motion.

Once someone has experienced vertigo, there is no question what it's like. And anyone who has experienced it agrees that once is enough! However, knowing what is happening can prevent it from happening.

Long-Term Weightlessness Effects

Although decades of human space experience have shown that a person can probably adapt to weightlessness and then readapt upon returning to Earth, all of the answers aren't known and won't be known until one or more generations of people live in space for most of their lives.

Soviet/Russian cosmonauts have spent longer periods of time in weightlessness than American astronauts. Space duration records have been broken continually, and this will go on.

Data from these long-duration stays in weightlessness indicate that there were measurable changes in body chemistry and function in such veteran space people. However, space medical experts believe that most of the physiological changes caused by weightlessness are reversible—i.e., once the person returns to Earth, the body changes caused by weightlessness disappear. But changes do occur in space.

Any part of the human body that isn't used on a regular basis will eventually decay or atrophy.

Bones and muscles begin to show changes after a person stays a short time in weightlessness. As might be expected in weightless living, bones and muscles don't have as much work to do supporting the body compared to what's re-

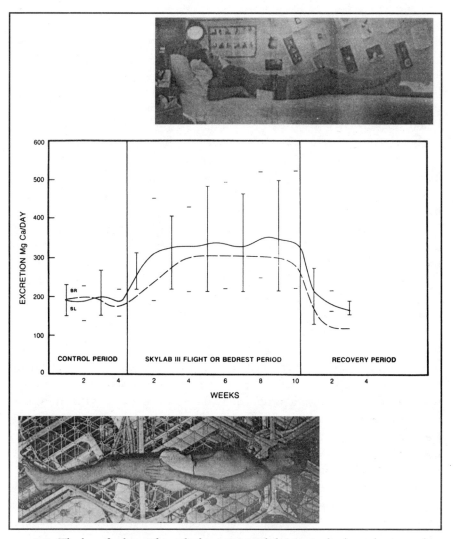

FIG. 6-6: *The loss of calcium from the bones in weightlessness is closely analogous to that lost on Earth during prolonged bed rest.* (NASA photo)

quired of them in Earth's gravity. After someone has been in weightlessness for several months, muscles lose strength and tone, causing some initial difficulties in moving around upon return to Earth.

But bone decalcification appears to be a significant problem connected with long-term living in weightlessness. Bones aren't required to be as strong in weightlessness and therefore begin to lose the calcium that contributes to their strength.

Bone decalcification in weightlessness occurs at a rate of between 1 and 2 percent of the total bone mass per month.

When bones lose calcium, it's released into the body fluids, primarily the blood serum because this is where calcium goes when a person ingests it in food. Ninety-nine percent of body calcium is located in the bones and one percent of this is freely exchangeable with extracellular body fluids. Normal blood calcium level is 8.8 to 10.4 milligrams per 100 milliliters of blood. Forty percent of this is chemically bound to serum blood proteins.

An excess of calcium in the blood is called *hypercalcemia* and occurs when blood calcium exceeds 10.5 milligrams per 100 milliliters. On Earth, hypercalcemia can be caused by excessive ingestion of food having high calcium content, by cancer, or by endocrine imbalances.

Symptoms of mild hypercalcemia include constipation, nausea, vomiting, abdominal pain, and often the urge for nearly constant urination. None of these symptoms have been reported by any astronaut or cosmonaut.

When the blood calcium level exceeds 12 milligrams per 100 milliliters, severe symptoms such as confusion, delirium, psychosis, stupor, and coma are possible. When it exceeds 18 milligrams per 100 milliliters, the results include shock, liver failure, and death. None of the severe hypercalcemic symptoms have been reported by astronauts or cosmonauts.

The mild forms of hypercalcemia that have been measured in astronauts have shown some imbalance of the blood electrolytes as well as imbalances in some hormone outputs and proportions. There are indications that these normal hormonal imbalances created unstable protein and carbohydrate states, hypoglycemia, and unusual increases in both primary and secondary pituitary hormone levels. These are the sort of symptoms doctors expect to see in cases of mild hypercalcemia here on Earth.

Hypercalcemia is a metabolic or mineral imbalance disorder that can be treated with immediate results in most cases by means of steroid therapy using a drug called prednisone. Long-term treatment of hypercalcemia involves the oral administration of phosphorus.

Whether or not the terrestrial therapy for hypercalcemia can be used to treat someone who has suffered from calcium resorption due to weightless living remains to be seen. Although it may be a matter of some concern for those who might spend a very long time in space, the fact that there is existing therapy for hypercalcemia bodes well for the eventual solution to the problem.

Cardiovascular Changes

Weightlessness has also caused changes in the heart and cardiovascular system of all astronauts and cosmonauts.

FIG. 6-7: *Human beings have had little trouble adapting to weightlessness and have actually enjoyed it.* (NASA photo)

On Earth, the heart and its system of blood vessels must contend with the problem of pumping blood up from the legs and feet while at the same time maintaining the proper blood pressure and flow to the brain on top of the body. It must also contend with the fact that people lie horizontally as well as stand erect.

Post-flight medical examinations of astronauts and cosmonauts upon their return to Earth has revealed an increase in heart rate by 10 to 20 beats per minute, changes in muscle reflexes, and pooling of blood in the lower abdomen and legs. Initial concerns that long-term living in weightlessness might lead to irreversible changes in the cardiovascular system coupled with a decreased effectiveness of the body's resistance to disease because of changes in blood chemistry have largely been proved to be unfounded as more people spent longer periods in space.

However, all of these physiological changes caused by weightlessness will be monitored for years to come. Therapies for combating them will undoubtedly be developed if they turn out to be serious.

Please keep in mind that all the data aren't in yet and won't be for decades. The space environment is new to humankind. People who travel and work in space will continue to be studied to provide this data.

And weightlessness is one of the things that astronauts, cosmonauts, and other space travelers look forward to experiencing. Said astronaut Dr. Joe Kerwin of the Skylab 2 crew, "It was a continuous and pleasant surprise to me to find out how easy it was to live in zero-g and how good we felt."

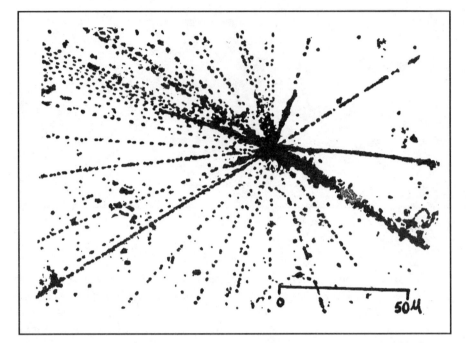

FIG. 7-1: *When high-energy ionizing radiation collides with atoms or molecules, fission or destruction of the molecule can occur.*

CHAPTER SEVEN

RADIATION

Space was once considered a quiet vacuum in which basic concentrations of matter known as stars, planets, planetoids, moons, comets, and meteors moved in predictable paths. Then in the twentieth century a new factor entered the picture: *ionizing radiation*.

The term "radiation" itself includes radio waves, microwaves, infrared, light, and ultraviolet waves. Ionizing radiation was discovered on Earth and comes from X rays, radium, and other radioactive sources. Early in the twentieth century, scientists detected "cosmic rays," a form of very energetic ionizing radiation that came from all directions in space.

Not all radiation is ionizing radiation and not all radiation is harmful to human beings. In fact, some radiation such as light is necessary and beneficial. Ionizing radiation, in general, is not. This chapter will explain why and what are the most detrimental forms of ionizing radiation for human beings on Earth and in space.

No one anticipated the space radiation that was detected by the first United States' satellite, Explorer-I, launched on January 30, 1958. The instruments in Explorer-I, designed by Dr. James Van Allen of the State University of Iowa, were intended to detect and measure the amounts of radiation in low-Earth orbit. The experiments were carefully designed using instruments similar to Geiger counters, whose measurement ranges were predicated on the basis of cosmic-ray experiments made over a few places on Earth—White Sands, New Mexico, and Fort Churchill, Canada, to name but two. These measurements had been made to altitudes of about 150,000 feet using balloons, and to 100-mile altitudes using rockets, but over shorter time. Explorer-I provided scientists with their first chance to gather cosmic-ray data from an experiment that would stay in space for months.

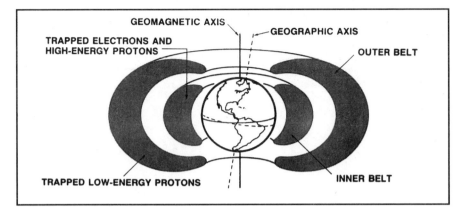

FIG. 7-2: *The Van Allen belts of trapped charged particles around the Earth.* (Art by Sternbach)

The Van Allen Belts

Explorer-I ran into a storm of ionizing radiation that overloaded Dr. Van Allen's instruments. He'd had no reason to believe such intense radiation existed in space. Dr. Van Allen redesigned the equipment using radiation counters with higher saturation limits. This was launched in Explorer-III on March 26, 1958, and reported the exact radiation level. Later satellite flights and space probes mapped what turned out to be belts of ionizing radiation that surround the Earth. These bear Dr. Van Allen's name today.

An understanding of the true nature of the Van Allen radiation belts is extremely important to people who live and work in space. Not only do the Van Allen belts offer people some protection against solar radiation if the facility is orbiting below the belts, but the presence of the belts themselves opens the extremely important subject of the effects of ionizing radiation on people in space.

The Van Allen belts exist because the Earth has a magnetic field and orbits in the atmosphere of a star, the Sun. The Earth is one of the few planets possessing a strong magnetic field. This field exists, according to the latest theories, because the Earth has a molten nickel-iron core. Eddies caused by the rotation of this molten core produce small magnetic fields. When the fields of all eddies are combined, they add up to a large terrestrial magnetic field with invisible lines of force that swoop out into space from the magnetic poles located near but not within the Earth's poles of rotation. This large magnetic field isn't neatly symmetrical like the field of a small bar magnet. It's distorted by the interacting magnetic field of the Sun and the charged particles blasted into space by solar activity.

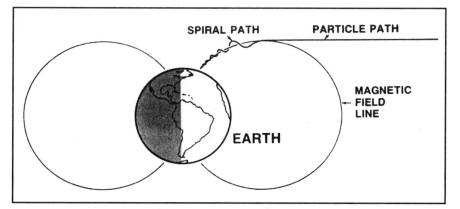

FIG. 7-3: *Incoming nuclear particles from the solar wind are captured by the Earth's magnetic field and spiral around the magnetic lines of force back and forth between the Earth's magnetic poles.* (Art by Sternbach)

The Solar Wind

The hydrogen-fusion energy process of the Sun produces charged subatomic particles—negatively charged electrons and positively charged protons. These stream out from the Sun as part of the "solar wind." During the Earth's journey through space around the Sun at an orbital speed of about 18 miles per second, it runs into enormous quantities of these charged subatomic particles ejected from the Sun.

Since electrons and protons have an electric charge, they're deflected from their original paths when they encounter the Earth's magnetic field. Some become trapped in the terrestrial magnetic field and circle back and forth between the Earth's magnetic poles along the invisible magnetic lines of force. Some of them escape again into space. Others encounter the Earth's upper atmosphere near the poles where, if the Sun is in an active period and putting out a lot of solar flare particles, these interact with the upper atmosphere to create the auroras.

This cloud of trapped solar particles forms doughnut-shaped belts around the Earth roughly over the equator. The density of the charged particles in the belts reaches several maximums from where the belts begin about 100 miles up to where they finally disappear about 17,000 miles above the Earth's surface.

An electrically charged subatomic particle can create ionizing radiation—mainly X rays or gamma rays—by a process called *bremsstrahlung,* a scientific German word meaning "radiation from slowing down" a particle. To understand bremsstrahlung, which is an important part of space living, let's take a visit to a local hospital.

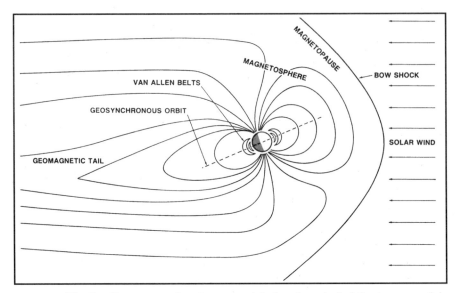

FIG. 7-4: *The Earth's magnetosphere interacts with the solar wind, deflecting most of the solar wind particles.* (Art by Sternbach)

X Rays

In 1895, Wilhelm Roentgen of Germany was experimenting with a glass vacuum bottle having electrodes at both ends. The device is called a Crookes tube because it was invented by Sir William Crookes in 1879. When Roentgen applied several thousand volts of electrical potential between the electrodes of the Crookes tube, he generated a new form of electromagnetic radiation that passed through human bodies. If a photographic plate was exposed to the radiation coming through the body part, the bones and other innards of the body thus irradiated could be seen on the film.

Today's hospital X-ray machine still uses a Crookes tube. One electrode contains a heated filament or cathode from which electrons are "boiled off" in the same manner as inside a television picture tube (yet another form of the Crookes tube). At the other end of the tube is a positive electrode, the anode, which is charged several thousand volts positive with respect to the heated cathode. The electrons leave the hot cathode and accelerate rapidly in the high-voltage electric field between the cathode and the anode, acquiring high energy as they travel. This energy is expressed as "electron volts."

An electron volt is defined as the amount of energy given to a particle with a unit electric charge when accelerated through a potential difference of one

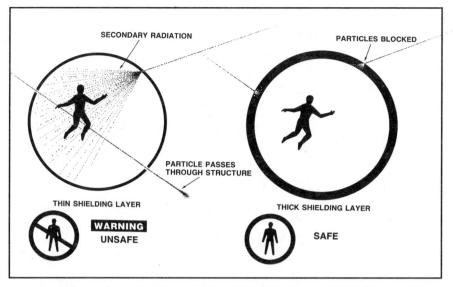

FIG. 7-5: *When a charged nuclear particle collides with the metal hull of a spaceship or space facility, it releases X rays by the* bremsstrahlung *phenomenon.* (Art by Sternbach)

volt. Because of the very small value of this unit, particle energies are usually expressed in multiples such as "million electron volts" or MeV.

The greater the voltage difference between the electrodes of a Crookes tube, the greater the electron speed, and, therefore, the higher the MeV energy. The electrons finally slam into the anode, where their energy is converted into heat and X rays that emanate from the tube.

Any electronic vacuum tube with several thousand volts across its electrodes will generate X rays. In fact, a television picture tube generates low-energy or "soft" X rays, but these are stopped by the leaded glass of the picture tube's screen.

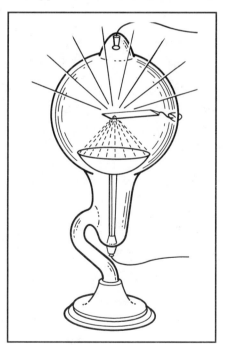

FIG. 7-6: *The original form of the Crookes tube.* (Art by Sternbach)

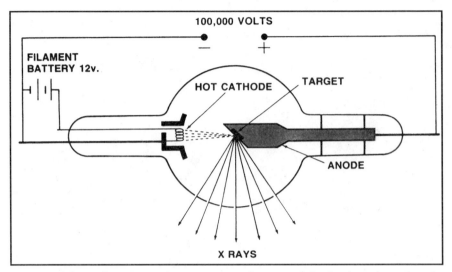

FIG. 7-7: *The modern X ray tube is an improved version of the Crookes tube and works the same way.* (Art by Sternbach)

Therefore, the same principle applies to any spaceship or facility traveling at high speed in space and encountering charged subatomic particles. The encounters will cause the generation of ionizing radiation because the facility collides with protons and electrons of the solar wind or those trapped in the Earth's Van Allen belts. When these charged particles slam into the metallic outer hull of a spaceship or facility, the collision duplicates the action inside a Crookes tube and radiation is produced by bremsstrahlung.

Bremsstrahlung radiation is only one form of ionizing radiation present in space. Other forms and sources of radiation in space can be deadly, too.

Other Ionizing Radiation

Ionizing radiation is of two sorts: (1) high-velocity, high-energy charged particles of the sort just discussed, and (2) high-energy uncharged photons in the electromagnetic spectrum.

As has been explained, particle radiation is primarily composed of either electrically charged or naturally charged subatomic particles—the negatively charged electron, the positively charged proton, and the neutral neutron. Particle radiation also may come from high-velocity atoms that have had their outer shell electrons stripped off and therefore carry an electric charge. Although the number of different kinds of nuclear and subnuclear particles keeps growing constantly as physicists probe the basic nature of matter and seek to understand it, the electron, proton, and neutron suffice to characterize the subatomic particles found in space.

The electromagnetic (e-m) spectrum is shown in Figure 7-8. E-m radiation is characterized by its frequency and wavelength—the two are directly related, the higher frequencies having the shorter wavelengths. Because e-m radiation can behave as if it's composed of uncharged particles of negligible mass, e-m radiation is considered to consist of packets of energy called "photon" particles. The higher the frequency of the photon, the greater the photon's energy expressed in MeV.

The low-energy, low-frequency end of the e-m spectrum contains the variations of the Earth's magnetic field known as the Schuman Resonance at six to ten cycles per second or *hertz* up through the radio waves used by AM and FM radio broadcast stations—from 0.5 to 1.6 million hertz (megahertz or mHz) for AM and from 88 to 108 mHz for FM. Above the radio frequency spectrum are other specialized radio frequencies and the microwaves that include those used for radar and for cooking food in domestic microwave ovens. These blend into the infrared or radiant heat portion of the e-m spectrum which, in turn, merges into the visible light portion that people detect as color, red being the longest light wavelength and violet being the shortest. Immediately above this is the ultraviolet portion, then the X ray segment that's capped by the gamma rays. On the upper end of the e-m spectrum are the cosmic rays that have the shortest wavelengths, highest frequencies, and highest-energy photons.

Human Radiation Sensors

Human beings are wonderfully equipped to sense what's happening around them. Evolutionary processes over the past several billion years have equipped life forms on Earth with sensing devices that help them survive, especially those that report danger. Humans are well-provided with natural sensors to detect pressure, sound, light, heat, acceleration, form, color, and so forth. These are shown in Figure 7-9 along with the sensations they produce and the interpretation people give to these sensations. Human sensors for the e-m spectrum are excellent for handling the survival factors with which human ancestors had to contend as hunters. However, their range and sensitivity are limited. With modern nuclear technology and the thrust into the new environment of space, these limitations can cause trouble because people can't

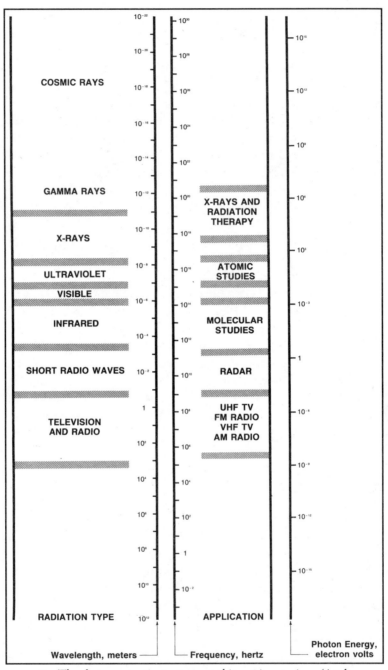

FIG. 7-8: *The electromagnetic spectrum and its various regions.* (Art by Sternbach)

Human Stimulus-Receptor-Sensation Relationship			
Stimulus	*Receptor*	*Sensation*	*Interpretation*
E-m waves, 10^{-5} to 10^{-4} cm	Photoreceptor	Light, color	All visual impressions
E-m waves, 10^{-4} to 10^{-2} cm	Skin thermo-receptors	Temperature hot to cold	Heat, fire, warmth, comfort, freezing
Mechanical oscillations 20-20,000 Hz	Inner ear, cochlea	Noises, sound	Sounds, voices, music tones
Pressure	Tangoreceptors of the skin and pressoreceptors in the body	Touch, weight, acceleration, deceleration	Objects, liquids, states of motion
Linear accelerations	Otoliths	Changes, state of motion, equilibrium	Movements, voluntary and involuntary motion
Angular accelerations	Semicircular canals	Rotation	Turning, voluntary and involuntary rotations
Chemicals in watery solution	Taste buds	Sweet, bitter, sour, salt	Food
Chemicals in gaseous state	Olfactory cells	Smells	Odors of certain substances
Chemical and mechanical inner changes	Proprioceptors of the muscles and connecting tissue	Internal tension and pressure	Hunger, thirst, digestive processes
High-energy effects of all kinds	Free nerve endings of sense organs	Pain	Extreme stress, injury, illness

FIG. 7-9.

sense some regions in the e-m spectrum that have definite effects on their physical health and well-being.

People have excellent sensors for light radiation: the eyes. Infrared can be detected by the skin. The results of over-exposure to ultraviolet can be detected if a person suffers a sunburn. But humans can't detect radiation in other parts of the e-m spectrum. If they could, anyone could tune in a radio station without using a transistor radio.

It wasn't known until the Second World War that human beings could see ultraviolet (u-v) radiation. Research in Great Britain indicated that a few airplane pilots could see u-v. Therefore, it was proposed that some airfields of the Royal Air Force be illuminated at night with u-v so these unusual pilots could see to land their bombers. However, it turned out that the type of people who could see u-v are blond, blue-eyed, Teutonic types—and the German Luftwaffe had plenty of pilots of that sort who also would be able to see the airfields!

Electromagnetic radiation of all types can have some effect upon a person. The physiological effects of some portions of the e-m spectrum are well-known—sunburn, for example. But the effects of some other types of e-m radiation aren't yet known or even understood. And a great deal is misunderstood and therefore feared.

Once misunderstanding and fear take root among people, it is terribly difficult to dislodge them even with scientifically conducted experiments, solid data, and facts. This is true of the feared physiological effects of e-m radiation in the range from 50 hertz to 500 mHz. Although there *may* be some long-term effects from high-level exposure to such e-m radiation, data indicate that everyday levels of exposure have no damaging effects. Some effects may manifest themselves in the range of 5 to 15 hertz where man-made e-m radiation may interfere with the "biological clocks" that are like the timing circuits in a computer; this extremely-low-frequency radiation may interfere with the 8.73 hertz variation in the Earth's magnetic field that has been termed the "Zeitgeber" or "time giver" when sensed by the human nervous system.

Microwaves—radio waves with frequencies above 1,000 mHz—at certain frequencies and power levels can be harmful to living beings because organisms contain large amounts of water. The water molecule is resonant to microwave energy at 2,450 mHz and will therefore absorb energy at that frequency. This is converted into molecular vibration that is manifested as heat. This is the way a microwave oven operates. Its microwave generator, usually a radar device called a magnetron, produces radiation that excites the water molecules of food and other substances placed inside its shielded enclosure.

But except for the examples cited, human beings are not equipped to detect e-m radiation. They must use external sensing devices they've developed.

The Physiological Effects of Radiation

At frequencies and wavelengths above the ultraviolet, photon energies are high enough to create damage in living organisms. Scientists, nuclear engineers, and technologists understand this process, one that is not well understood by people in general. Since people can't sense high-energy radiation in the ultraviolet, X ray, and gamma ray portion of the e-m spectrum, they have a tendency to harbor an unreasoning fear of it. The word "radiation" itself has a highly negative semantic content—it invokes fear and apprehension. In this book, "radiation" has been used in its precise scientific context. There is nothing to fear from radiation per se, and some types of radiation are actually beneficial to human beings.

The *energy* content of radiation is the factor that can harm living organisms.

When a photon with radio wave MeV energy goes through a living cell (as uncountable numbers of them do every day), it creates no damage because it doesn't have enough energy to do so. At infrared frequencies, it excites all atoms and molecules, creating heat. At light wavelengths, its energy can be chemically detected by special, highly evolved cells in the retina of the eye. However, a light-energy photon doesn't do any damage unless a large number of them impinge on a small area in a short time as happens with coherent light from a laser.

However, when the energy level of a photon classifies it in the ultraviolet and X ray portion of the e-m spectrum, the photon can indeed cause microscopic damage inside a living cell.

When a high-energy photon plows into a living cell, the matter in the cell slows it down and makes it give up some of its energy. The photon transfers energy by ionizing some of the atoms in the cell, stripping them of their orbital electrons and leaving them with an electric charge. These ionized atoms don't behave the same in a chemical sense as their non-ionized companions. Their presence disrupts the delicate atomic and molecular balances within the cell, changing it so that it doesn't work the same—or as well—as before.

Measuring Radiation

Much of what has been written about ionizing radiation emphasizes the hazards and uses scare tactics. Perhaps some rationality can be introduced into the subject by recalling the words of Lord Kelvin quoted earlier concerning the importance of numerical measurement in scientific inquiry.

The various types of radiation can be detected and measured. The relative biological damage caused, if any, also can be measured and is known.

The most common measure of radiation dosage is the *rad*, which stands for "radiation absorbed dose." This is the amount of energy dissipated by ionizing radiation in matter. One rad equals 0.0000024 calories per gram. This isn't very much.

A similar unit called the *roentgen*, named after the nineteenth-century scientist who discovered X rays, is also a measure of exposure dose. One roentgen of gamma radiation produces 87 ergs of energy per gram of mass. An *erg* is the amount of energy required to move one gram of mass one centimeter—again, not a large value.

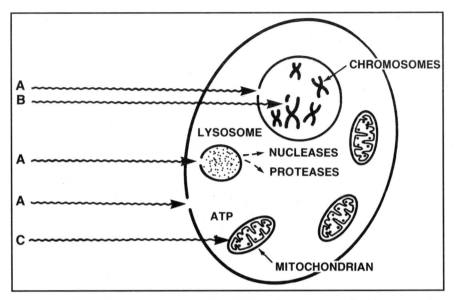

FIG. 7-10: *When charged nuclear particles collide with an organic cell, the energy released strips away electrons from the cell's molecules, ionizing them and changing their chemical characteristics.* (Art by Sternbach)

Another unit called the *rem*—an acronym for "rad equivalent, mammal"—is a unit reflecting the ability of ionizing radiation to produce damage in the living tissue of a mammalian cell.

The *dose rate* is the radiation received per unit time. Detectable biological effects increase as the dose rate increases. Roughly stated, an observable effect is certain with dose rates greater than 4 rads per minute.

Body area exposed is also a factor. The entire human body can absorb up to 200 rads without a fatal outcome. However, as the whole-body dose approaches 450 rads, the death rate becomes about 50 percent. A whole-body dose of more than 600 rads in a short time is always fatal. But many thousands of rads can be delivered to a human body for cancer therapy over a long period of time if only small portions of the body are irradiated. If the bowels and the bone marrow are adequately protected by shielding, an individual may survive what might otherwise be a fatal whole-body dose.

Ionizing radiation is measured using several different instruments. The most common of these is the *Geiger counter*, a device that can perhaps be described as a near-vacuum tube. The main sensor is a tube made of metal with an insulated wire running down the central axis. When a particle—high-energy photon or charged particle—enters the tube, it produces an ionized pathway through the near-vacuum. This permits a brief electrical current to pass between the central wire and the case. Suitable electronics then can be used to detect and count the number of short-outs in a given time. Sometimes this output is audio, producing a click every time a particle transits the tube.

Another instrument is the photographic *dosage badge*, a small piece of photographic film that is worn as a badge by a person. After a given time, the film is developed and the amount of radiation it has absorbed is determined by the amount of fogging of the emulsion. Other types of radiation sensors exist such as one that uses photoelectric cells to detect the flashes produced when a charged particle impacts a phosphor layer inside the sensor. Yet others use the principle of the electroscope, that magical device that looks like a light bulb with a small piece of material inside that looks like tin foil.

Natural Radiation Dose

A gentle, low-intensity rain of high-energy radiation occurs all the time on Earth. It comes from the Earth itself as well as from space. The Earth-source radiation comes from naturally radioactive elements that are present in nearly all rocks and soils. The space-source radiation comes from the Sun and from the galaxy. Organisms on Earth have adapted to living under this continual bombardment of natural radiation; if it were taken away somehow in an attempt to create a "pure" environment without any "bad radiation," no one really knows what the long-term effects on the Earth's living organisms would be.

The cells of the human body are subjected also to this natural or "background" radiation. Human cells die and replace themselves regularly. Because of this, a person today isn't the same as last year. Human beings have developed an immunity to background radiation.

Each year, the average human being is subjected to the following dose rates in millirads per year (mr/yr):

Natural radioactive material in the bones:	34 mr/yr
Cosmic rays (at sea level):	30 mr/yr
Natural radioactivity of the surroundings:	48 mr/yr
Medical X rays:	75 mr/yr
Radioactivity from man-made sources:	about 12 mr/yr
TOTAL:	about 200 mr/yr average

A person can handle this natural background radioactivity without any discernable physiological effects. In fact, the development of life on Earth may have depended upon this background radiation to create genetic muta-

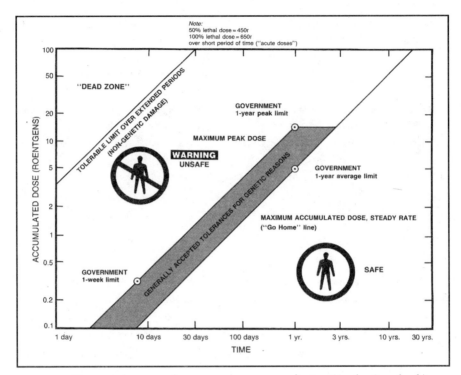

FIG. 7-II: *Human tolerance to long-term exposure to radiation.* (Art by Sternbach)

tions, which in turn caused the variation as well as the development of species such as *Homo sapiens.*

According to the National Academy of Sciences, a person can absorb up to 10 rads in a thirty-year period without danger, including about 6 rads from background radiation in that time period. Human physiology can handle more than that. Nuclear workers are permitted to receive dosages of 300 millirads per week for a total of 46.8 rads in a thirty-year period.

Figure 7-13 shows the operational exposure limits in rems permissible for NASA Space Shuttle flight crews in space.

Heavy Radiation Exposure

If a person gets too much high-energy ionizing radiation, so many cells may be affected that they can't replace themselves. Or the radiation may affect cells so that their natural functions are disrupted, turning them into rampaging, ever-growing, all-consuming cancer cells.

If too many cells are damaged, radiation sickness is the result.

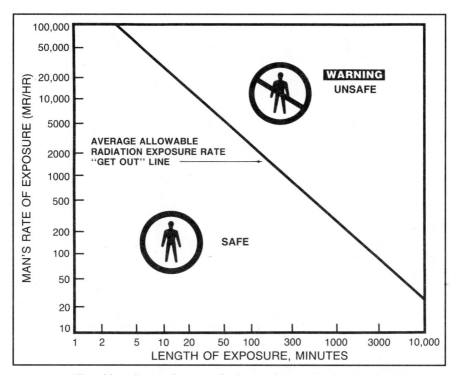

FIG. 7-12: *Allowable radiation dose rates for human beings.* (Art by Sternbach)

The physiological effects of ionizing radiation are well known. Body tissues vary in their response to immediate radiation in the following order of sensitivity, the most sensitive listed first:

1. lymphoid cells
2. reproductive cells of the gonads
3. bone marrow cells
4. epithelial cells of the intestines
5. the epidermis (skin)
6. hepatic (liver) cells
7. the epithelium (lining) of the lungs
8. kidney tissue
9. cells of the intestinal cavity
10. nerve cells
11. bone cells
12. muscle cells and connective tissues

NASA Operational Flight-Crew Exposure Limits

(All exposure units expressed in rems)

Constraint	Bone Marrow	Skin	Eye
Thirty-day maximum	25	75	37
Quarterly maximum	35	105	52
Yearly maximum	75	225	112
Career limit	400	1,200	600

FIG. 7-13

Large but sublethal doses affect both the rate of mitosis (cell division) and the synthesis of DNA. This can lead to diminished production of new cells in tissues that normally undergo continual renewal—bone marrow and the gonads, for example. Some cells can be so badly damaged by ionizing radiation that they may continue to divide but produce abnormal progeny. Some of these altered cells may be cancerous.

A rule of thumb for exposure to ionizing radiation is: a little is probably necessary but a lot over a short time isn't. A lot of it over a long time may or may not be harmful, depending where it was absorbed by the body.

Radiation in Space

Although a human being can handle the natural or background radiation on Earth—and to a large extent even the unnatural radiation from nuclear powerplants and other man-made sources if the dose rates are low—measurements made of the type and amount of radiation in space indicate that this unseen hazard definitely poses a problem for unprotected humans in space.

Figure 7-14 shows the types of ionizing radiations that exist in space above the Van Allen belts, their energy levels in MeV, the "relative biological effectiveness" (known as RBE) of each, and where each type comes from.

Because different forms of ionizing radiation with different energies may have different effects upon organic tissue, the radiation dose from each must be multiplied by the RBE. The product then is a measure of the danger of the particular kind of radiation and is usually presented in rems. Thus, a low-

Ionizing Radiations in Space		
Name	*RBE*	*Source*
X rays and gamma rays	1	Radiation belts, solar radiation, and bremsstrahlung electrons
Electrons		
1.0 MeV	1	Radiation belts
0.1 MeV	1.08	
Protons		
100 MeV	1–2	Cosmic rays, inner-radiation belts, and solar cosmic
1.5 MeV	8.5	rays
0.1 MeV	10	
Neutrons		
0.05 eV		
(thermal)	2.8	Nuclear interactions in the sun
0.0001 MeV	2.2	
0.005 MeV	2.4	
0.02 MeV	5	
0.5 MeV	10.2	
1.0 MeV	10.5	
10.0 MeV	6.4	
Alpha particles		
5.0 MeV	15	Cosmic rays
1.0 MeV	20	
Heavy primaries	*	Cosmic rays

*Damaging power of heavy primaries varies widely and is measured in terms of how many chemical bonds per unit of body mass are broken, thereby giving an indication of the tissue damage sustained.

(From "Space Settlements, A Design Study," National Aeronautics and Space Administration publication SP-413, Government Printing Office, 1977.)

FIG. 7-14

energy thermal neutron would have 2.8 times the effect of a one-rad exposure to X rays.

In the Van Allen belts, X rays and gamma rays are generated by bremsstrahlung electrons and protons as well as from nuclear reactions.

Electrons come from the Sun in the solar wind and are the most numerous charged particles trapped in the Van Allen belts. Their primary hazard is their ability to produce bremsstrahlung radiation in the X ray and gamma ray regions of the e-m spectrum when they collide with the metallic hull of a spaceship or facility.

Protons also come from the Sun in the solar wind and are trapped in the Van Allen belts. Protons are also part of the overall high-energy radiation called cosmic rays.

Neutrons are produced by nuclear interactions. Their primary natural source in the solar system is the thermonuclear fusion process of the Sun.

Alpha particles are part of cosmic rays and, even at low energies, have large RBE values.

The "heavy primaries" listed in Figure 7-14 are the major constituents of cosmic rays.

Cosmic Radiation

Cosmic radiation has been a source of much frustration to nuclear physicists and appears to be composed mostly of high-energy atomic nuclei—i.e., atoms that have had all their orbital electrons stripped away. About 80 percent are hydrogen nuclei or protons; 16 percent are helium nuclei consisting of two protons and two neutrons with an electric charge of +2; and 4 percent are made up of the nuclei of heavy atoms such as iron. Cosmic-ray energies are extremely high, ranging from 100 MeV to several thousand MeV. The source of cosmic rays is unknown, although some probably originate in the Sun while the remainder may be from elsewhere in the galaxy because they come from all directions in space.

The Earth's atmosphere acts as an excellent shield. Only a small percentage of cosmic radiation reaches the ground. The atmosphere's effect on cosmic radiation is somewhat like that of a dense forest on a rifle bullet: the bullet (cosmic ray) won't go very far before it hits a tree (atmospheric molecule). The heavier cosmic-ray primaries collide with atmospheric molecules at high altitudes while the lighter, smaller, less energetic particles penetrate farther into the atmosphere before they suffer a collision.

When a cosmic-ray particle, especially one of the slow-moving heavy primaries, hits an atmospheric molecule, the energy is released in a nuclear disintegration with protons, electrons, neutrons, gamma rays, and other nuclear debris scattering in all directions. The spray of nuclear junk resulting from the collision of a heavy cosmic-ray primary with an atmospheric molecule is called "secondary cosmic radiation." The energy of these secondaries is extremely high within several millimeters of the occurrence.

On the surface of the Earth, secondary cosmic radiation is everywhere. However, it generally isn't harmful to humans because a cosmic-ray particle must have an energy of more than 1,000 MeV to reach the ground at the Earth's equator. Secondary cosmic radiation is observed with increasing intensity as altitude increases. However, at altitudes above 75,000 feet, only primary cosmic-ray particles exist because atmospheric density isn't high enough to produce collisions and thus secondaries.

The light, fast, and most common cosmic-ray particles are the protons (hydrogen nuclei or ionized hydrogen) and alpha particles (helium nuclei or

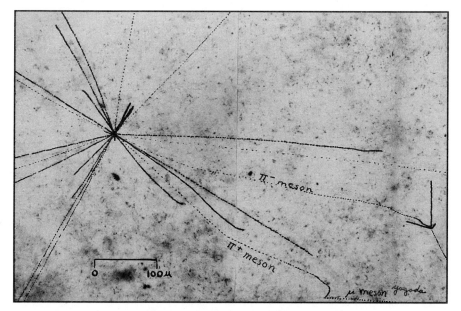

FIG. 7-15: *When nuclear particles collide with atoms and molecules, they create a spray of "secondary" particles or radiation as shown in this historic photo of a cosmic ray collision made by an unmanned rocket cosmic ray detector of Dr. Herman Yagoda in 1955.* (Dr. Herman Yagoda)

ionized helium). They have up to 20 times the ionizing potential of X rays, but they have a tendency to pass through organic tissue so rapidly that there is little time to cause ionization. The slow, heavy primaries are the most destructive to organic tissue because they leave a broad trail of ionization that kills cells. When it comes to cosmic radiation, the ionizing power increases as the particle energy decreases, within limits.

Solar Flares

Figure 7-14 doesn't indicate the relative population or concentration of ionizing radiation in space because this depends on where the measurement is taken as well as what's happening on the Sun at that moment.

The Van Allen belts trap many of the solar wind electrons and protons while the full force of the solar wind exists above the Van Allen belts.

During an event known as a solar flare, the concentration of charged particles from the Sun as a result of the flare's energy release can increase enormously, producing a high-radiation environment above the Van Allen belts that is lethal to human beings.

All of the manned space flights to date—except the Apollo lunar landing flights—orbited below the Van Allen belts and therefore enjoyed protection against radiation. To ensure the safety of the Apollo astronauts, the flights to the Moon not only went through the Van Allen belts very rapidly—to prevent extensive exposure of the astronauts to the charged particles there—but were also scheduled during periods when the Sun was relatively quiet and no large solar flares were anticipated.

Protecting People Against Radiation

All e-m radiation and especially its ionizing forms can be stopped by varying thicknesses of material or mass. X rays can be stopped by a few inches of lead. Alpha particles can be stopped by a sheet of paper. The intensity of high-energy ionizing radiation such as gamma rays decreases rapidly as the total thickness of mass through which they must pass increases. These are the types of ionizing radiation that are easy to stop.

Both primary and secondary cosmic rays are difficult to stop because of the high energies involved. And when primaries plow into matter, they produce secondaries that can be made up of very energetic charged particles as well as gamma rays and X rays.

Neutrons are difficult to stop because they're electrically neutral and have no electrical charge. Some elements such as beryllium will absorb neutrons.

Because of bremsstrahlung radiation, the best way to protect humans against ionizing radiation in space is to have a spaceship or facility just thick enough to keep air in and ultraviolet radiation out.

For long-duration flights and facilities above the Van Allen belts, "storm cellars" will have to be included in the designs. These would be modules with very thick walls to shield occupants against the intense but brief radiation produced by solar flares.

However, in 1986 the Department of Defense revealed their research results on "hardening" humans against ionizing radiation.

Nuclear explosions produce nuclear radiation that can destroy or damage electronic circuits as well as human beings. The armed services conducted research on how to protect or "harden" electrical and electronic devices against radiation. By 1985, they had produced a 64-bit computer memory chip that could withstand more than a million rads of ionizing radiation. However, they also supported research on "hardening" human beings, too.

In an October 1986 symposium, the director of the Defense Nuclear Agency announced research results showing that the human casualties from the ionizing radiation of nuclear fallout could be reduced to less than 10 percent.

According to the DNA, it's possible to "harden" a human being by several means.

The "prophylactic" measures—ones taken ahead of time—start with dietary supplements rich in green vegetables as well as massive doses of Vitamin A (retinol) and Vitamin E (tocopherol). Both are fat-soluble antioxidants found in green, leafy vegetables. Together, these two vitamins produce a 30 percent reduction in the effects of ionizing radiation in the cell.

Short-term measures taken at the time of radiation exposure include injections of atropine, a plant alkaloid which is an anticholinergic drug that opposes the actions of acetylcholine at nerve endings. This can reduce radiation effects by another 10 percent.

Post-exposure therapy includes drugs and various herbal preparations for intravenous feeding, purging of toxic chemicals from the cells, and repair of damage to the immune systems to restore the body's resistance to infection.

While ionizing radiation remains a problem for people in space in spite of recent and on-going research, it's not an insoluble problem that will keep people from living and working in space.

FIG. 8-1: *NASA space shuttle astronaut Dr. Rhea Seddon assembles food items in a portable warmer on the space shuttle Orbiter's mid-deck.* (NASA photo)

CHAPTER EIGHT

NUTRITION AND SANITATION

Nutrition and Sanitation on Earth

In common with an atmosphere and gravity, food is something people take for granted unless they don't have any and become hungry. Food is usually something created by someone else and obtained at a supermarket. The economic food chain is now so long and complex in high-technology Earth cultures that most people have only the most rudimentary knowledge of food technology. In the United States and most Western nations, malnutrition and its associated diseases and syndromes are seldom seen, even during the worst economic periods. On the other hand, overeating and its immediate and long-term consequences—indigestion and obesity—are common denominators of the high-technology cultures of Earth.

A different attitude prevails about sanitation. This attitude is exemplified by the NASA euphemism for it: "waste management." In high-technology cultures, the normal physical acts of urinating and defecating aren't subjects for open social discussion. When circumstances require talking about them, people use latinized words and euphemisms in place of the Anglo-Saxon or vernacular counterparts. While the physical act of eating is a social occasion, waste elimination is a private one. Almost the opposite is true in some other cultures of the world where people are closer to the problems of raw survival. A clear example of this can be seen in the difference between the two types of restrooms that have been built for tourists at the Great Wall of China. The "Western" bathroom has the usual ceramic equipment in private stalls while the Chinese bathroom has only a series of holes in the floor. In the Western world, sewage that includes human waste is treated to create a dried sludge that may be used as fertilizer. In the Orient, many privies are built over pigstys because swine do an outstanding job of converting human wastes into edible protein.

In nearly all situations on Earth, the waste problem has been solved by recycling it back into the huge closed ecological system of the planet so that the wastes don't cause disease or offend people with their odor.

Nutrition and Sanitation in Space

In space, however, nutrition and sanitation take on a new importance because the ecosystem of a spaceship or facility is much smaller, resembling those on long-range jet airliners. All the food for the flight is put aboard before departure. The waste is collected in bins or "honey tanks" aboard the airplane. This is later transferred to a "honey wagon" at the destination airport where the waste is put into an ordinary sewage treatment system. It's an "open system" that depends upon larger food and sanitation systems at the points of departure and destination. It's also a system that is time-dependent (it solves the nutrition and sanitation problems for a few hours). It's also people-limited (it will handle the problem for up to 500 people). Long-duration space facilities such as space stations can continue to use similar open systems if supply ships can bring food up from Earth and return the wastes to the surface regularly. Such systems can be used for a few hundred people living in space. Beyond that number, however, "closed ecological life support systems" must be used that are smaller versions of the Earth's own planetary closed system.

Because of the accumulated knowledge of aircraft practice and the limited experience gained from the NASA space shuttle Orbiter and early space stations such as Skylab, *Salyut*, and *Mir*, as well as from experimental closed systems that have been built and operated on Earth since 1957, a firm foundation exists for engineering closed nutrition/sanitation systems in spaceships and facilities.

Nutritional Requirements

As mentioned in an earlier chapter, the human body can be considered as a heat engine that converts organic fuel into heat energy and motion, utilizing the oxygen of the air for the metabolic process. Not all of this energy goes into useful work. A human being has an efficiency of about 11 percent. The remaining energy is waste in the form of heat exchanged with the environment as well as solid and liquid waste material.

Human metabolism is the sum of the processes by which the human body uses nutrients and oxidants. Two major variables control human metabolism. Because humans are different from one another, most of this difference is caused by the first variable: genetic inheritance. This determines the nature of the structure and function of the body's cells and therefore controls the individual's ability to utilize the food, water, and air from the environment. The second variable is the sort of nutritional material consumed. Two other minor variables are (1) intermediary metabolism, which is the way the body's metabolic processes interact based on chemical and physical properties of the environment and (2) the body's own regulation of the cellular and organ interrelationships—especially the endocrine system—that integrates tissue activities in a way conducive to continued health.

Although a person's energy requirements therefore vary from those of other people, the average adult human body requires between 1,800 and 3,600 kilocalories, or nutritional Calories, per day. The minimum acceptable caloric intake is about 1,600 Calories per day on a long-term basis.

But these "normal" caloric intakes depend upon body mass, sex, age, state of health, and level of physical activity. The normal daily caloric requirement for male adults is 21 Calories per pound of desired body weight while the adult female requirement is 18 Calories per pound. Young, growing humans often require as much as twice the normal daily caloric intake, as any parent of teenagers will attest. When people grow older, they tend to eat less. People who are ill often eat less as well. A nutritionist once remarked that the only sure-fire way to lose weight is to get old or sick.

Water Requirements

Although the human body produces and exhales water as one of the products of cellular combustion, it still requires about two quarts of water per day as either drinking water or as part of food. During periods of heavy physical activity that includes profuse sweating or in high-temperature environments, liquid intake requirements may increase to as much as six quarts per day.

The human body's need for water is second only to its need for a sufficient supply of oxygen. People have survived without water for up to seven days. However, severe dehydration can begin within a few days, depending upon the individual and the environment.

The primary effect of dehydration is an imbalance in the body's acid-base relationship, reduction in the amount of intercellular fluid, and major changes in the electrolyte balance of the blood. These produce conditions that, depending upon the relative severity of dehydration in a given individual, can produce other disorders.

Water may not be a problem in a spaceship or facility. An operable water-recovery system properly maintained and sanitized will be able to provide more than enough fresh, clean, potable water because, as discussed earlier, the human body produces water as a by-product of cellular combustion or metabolism. Handling excess water in the environment of space living may be more of a problem than getting enough of it.

Balanced Nutritional Requirements

Quantity and caloric content of food isn't enough for a human being who must also have a "balanced diet."

A typical balanced diet consists of 1,600 to 2,000 Calories of carbohydrates, 630 to 1,000 Calories of fat, 400 to 600 Calories of protein, and an adequate supply of both vitamins and minerals, regardless of what the current dietary fad might happen to be.

A large amount of information is available regarding diets, foods, etc. All sorts of "special" diets exist. Because of the nature of nutrition and sanitation during the early decades of space living, people won't have to be dieticians to eat properly in space. However, they should be aware of nutritional requirements because available space food may not have the variety available on Earth.

A constant watch must be maintained against "micronutritional" deficiency diseases. These are caused by lack of vitamins and minerals in a diet. Very small amounts of these micronutrients (0.00002 percent to 0.0005 percent of body weight) are required. Most but not all native diets on Earth have evolved around locally available foodstuffs that possess adequate vitamins and minerals to satisfy human needs.

Vitamin deficiency diseases are rarely encountered in North America and Europe because of the general abundance of most types of foodstuffs even during the depth of the harshest winter. Frozen vegetables, concentrated citrus juice, eggs, dehydrated milk, and other components of a balanced diet are shipped worldwide by air cargo. Thus, a breakfast in Fairbanks, Alaska, in December can start with orange juice and a business lunch in Manhattan during the same month can include a crisp tossed salad. Therefore, doctors rarely encounter vitamin deficiency diseases even in the most difficult economic times. Such wasn't always the case in the past. However, some circumstances in space could cause the flow of foodstuffs to be interrupted.

The current listing of daily requirements of micronutrients is shown in Figure 8-2.

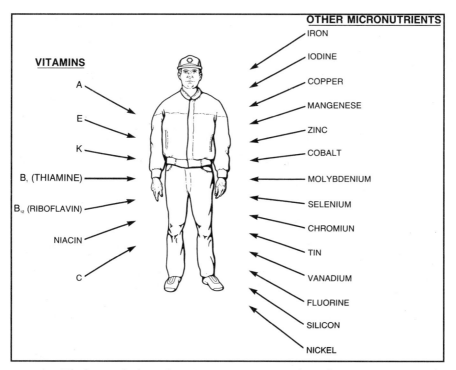

VITAMINS

A
E
K
B₁ (THIAMINE)
B₁₂ (RIBOFLAVIN)
NIACIN
C

OTHER MICRONUTRIENTS

IRON
IODINE
COPPER
MANGENESE
ZINC
COBALT
MOLYBDENIUM
SELENIUM
CHROMIUN
TIN
VANADIUM
FLUORINE
SILICON
NICKEL

FIG. 8-2: *The human body needs various vitamins, minerals, and micronutrients in order to remain healthy.* (Art by Sternbach)

Micronutrient Deficiency Diseases

Micronutrient deficiency symptoms should be known and understood by people living in space because some of them can be deadly.

The two types of vitamins are (1) fat soluble, and (2) water soluble. Fat-soluble vitamins can be stored in the human body while any excess of water-soluble vitamins is eliminated by urination or defecation.

Vitamin A (retinol) is a fat-soluble vitamin found in fish liver oils, liver, egg yolks, butter, cream, and vegetables. Vitamin A is readily and quickly destroyed by exposure to air and light. Most of the body's vitamin A is stored in the liver. Vitamin A deficiency primarily affects the eyes, leading to a deterioration of rod vision. Thus, a lack of vitamin A causes night blindness or an inability of the eyes to adapt to low light levels. In its advanced stages, it can cause drying, thickening, or wrinkling of the white portion of the eye, the conjunctiva.

Vitamin D is also fat soluble and is found in yeast, fish liver oils, and egg yolks. It's also formed when the human skin is exposed to ultraviolet light. It's stored in the kidneys. Lack of vitamin D causes rickets, a disease that decalcifies bones and thus creates deformities in growing children. Although vitamin D deficiency isn't a serious problem in space, the importance of vitamin D in calcium metabolism makes it critical because of the calcium resorption syndrome of weightlessness. Vitamin D plays a major role in calcium transport in the intestines and bones. The importance of vitamin D levels in the human body with respect to the hypercalcemic danger of weightlessness has yet to be evaluated.

The role of vitamin E (tocopherol) in human metabolism has not been thoroughly determined. This fat-soluble vitamin, present in vegetable oils and wheat germ, is an intracellular antioxidant—that is, it may retard or prevent some of the aging process in cells by maintaining the stability of cellular membranes.

Vitamin K, a fat-soluble compound present in green leafy vegetables, plays a role in the ability of the blood to clot and thus prevent hemorrhages.

One of the most important water-soluble micronutrients is vitamin B_1 or thiamine. Although thiamine deficiency in its extreme form produces the disease called beriberi, in its less severe form it manifests itself in failure of the appetite and in certain forms of neuritis. Thiamine occurs abundantly in cereal grains and the husks of unboiled rice. Beriberi is primarily seen among people who subsist on rice whose hulls have been removed by milling; the husk contains most of the thiamine, but boiling the rice before husking disperses the vitamin throughout the grain, thus preventing its loss.

Vitamin B_2 is a water-soluble alcohol called riboflavin that occurs in milk whey and egg whites. Lack of riboflavin leads to lesions or open wounds of the skin or tongue.

The vitamin that isn't called a vitamin, niacin or nicotinic acid, is classed as a water-soluble B vitamin found in most protein-rich foods such as lean meat. Severe niacin deficiency causes pellagra, which usually occurs where corn is a major part of the local diet and where corn has not been previously alkali-treated as it is in Mexican tortillas. Pellagra appears initially as lesions of the mouth, changes in the mucous membranes, burning sensations in the mouth, abdominal discomfort, nausea, vomiting, and diarrhea. Pellagra can cause organic psychoses, excitement, depression, mania, delirium, and paranoia. Because of the hazards of space living that could be caused by pellagra, diets should be monitored to prevent micronutritional deficiency.

Homo sapiens is one of the few animal species that cannot synthesize vitamin C (ascorbic acid) within the body. In comparison to other micronutrients, large amounts of vitamin C are required—70 milligrams or more per day. Citrus fruits as well as fresh vegetables and fruit are the best dietary sources of vitamin C. However, as a water-soluble vitamin, it is readily destroyed by oxidation. Any excess intake of vitamin C is passed through the body quickly. Early voyagers across Earth's oceans ran into the consequences of vitamin C deficiency: scurvy. This disease is characterized by weakness, lassitude, irritability, weight loss, little hemorrhages under the ends of the fingernails, and gross changes in the gums, which become swollen, purple, spongy, and bleed easily. Gangrene and loosening of the teeth may occur. Vitamin C can also retard various aging processes and has been shown in some cases that it can prevent cancer as well as the common cold.

Noncritical Micronutritional Deficiencies

Other micronutritional deficiencies aren't as critical or serious. Some of these cannot be detected or diagnosed except by laboratory tests. Diagnosis and treatment should be left to the medical department of a space facility because they can be detected during regular physical checkups.

Other micronutritional factors are the trace elements the human body needs. Currently, fourteen of these trace elements are recognized as essential for warm-blooded animals. They occur in concentrations of less than 0.005 percent of the body weight. In order of demonstrated importance they are iron, iodine, copper, manganese, zinc, cobalt, molybdenum, selenium, chromium, tin, vanadium, fluorine, silicon, and nickel. All are poorly absorbed and secreted from the human body and may therefore accumulate to the level of toxicity. Regular medical examinations should detect early signs of toxicity, which may be far more serious in space living than deficiencies of these trace elements.

Except for iron and iodine, it is generally uncommon for trace-element deficiencies to develop, no matter what sort of diet is being consumed. Most deficiencies are caused by genetic factors and are manifested in metabolic disorders. Most of these will be detected in preliminary physical examinations on Earth before an individual ventures into space, although some genetically caused trace-element deficiencies may only appear once family life becomes established in space facilities.

Space Food

A varied diet is just as important in space as it is on Earth. Not only is it necessary for food to possess the required caloric and micronutritional content, but it must also supply the necessary roughage to provide for the human need for fiber in the diet.

Space Shuttle Menu Design

The Shuttle menu is designed to provide nutrition and energy requirements essential for good health and effective performance with safe, highly acceptable foods. In order to maintain good nutrition, the menu will provide at least the following quantities of each nutrient each day:

Protein	(g)	56	Vitamin B_{12}	(g)	3.0
Vitamin A	(iu)	5000	Calcium	(mg)	800
Vitamin D	(iu)	400	Phosphorous	(mg)	800
Vitamin E	(iu)	15	Iodine	(μg)	130
Ascorbic Acid	(mg)	45	Iron	(mg)	18
Folacin	(μg)	400	Magnesium	(mg)	350
Niacin	(mg)	18	Zinc	(mg)	15
Riboflavin	(mg)	1.6	Potassium	(meq)	70
Thiamine	(mg)	1.4	Sodium	(meq)	150
Vitamin B_6	(mg)	2.0			

FIG. 8-3

Acceptability of food is just as important as balance. It's possible to have available sufficient food for a balanced diet and yet have people refuse to eat it.

To a large extent, food acceptability is primarily a cultural matter and secondarily a religious concern. While taste and appearance are factors, they seem to be of less importance. People simply won't eat certain foods, no matter how hungry they become.

Gas-producing foods should not be served to people in space who may be working in modules or space suits with reduced atmospheric pressures. Gas-producing foods can cause abdominal pains from gastric extension in addition to considerable flatulence, which may overload the air system purification equipment.

Food and Morale

Food is always a major morale factor. If the food isn't acceptable, morale deteriorates and performance drops. People may even become nauseated and unable to eat what food there is. Variety is a major acceptability criterion. Any menu plan duplicated in less than fourteen days becomes boring and therefore unacceptable, according to the decades of experience of the United States Navy, gained from feeding the crews of missile-launching submarines that remain submerged on station for as long as six months at a time.

Space Shuttle Food and Beverage List

Foods*

Applesauce (T)	Chicken and noodles (R)	Peach ambrosia (R)
Apricots, dried (IM)	Chicken and rice (R)	Peaches, dried (IM)
Asparagus (R)	Chili mac w/beef (R)	Peaches, (T)
Bananas (FD)	Cookies, pecan (NF)	Peanut butter
Beef almondine (R)	Cookies, shortbread (NF)	Pears (FD)
Beef, corned (I) (T)	Crackers, graham (NF)	Pears (T)
Beef and gravy (T)	Eggs, scrambled (R)	Peas w/butter sauce (R)
Beef, ground w/pickle sauce (T)	Food bar, almond crunch (NF)	Pineapple, crushed (T)
Beef jerky (IM)	Food bar, chocolate chip (NF)	Pudding, butterscotch (T)
Beef patty (R)	Food bar, granola (NF)	Pudding, chocolate (R) (T)
Beef, slices w/barbeque sauce (T)	Food bar, granola/raisin (NF)	Pudding, lemon (T)
Beef steak (I) (T)	Food bar, peanut butter/	Pudding, vanilla (R) (T)
Beef stroganoff w/noodles (R)	granola (NF)	Rice pilaf (R)
Bread, seedless rye (I) (NF)	Frankfurters (Vienna sausage) (T)	Salmon (T)
Broccoli au gratin (R)	Fruitcake (NF)	Sausage patty (R)
Breakfast roll (I) (NF)	Fruit cocktail (T)	Shrimp creole (R)
Candy, Life Savers‡, assorted	Green beans, french	Shrimp cocktail (R)
flavors (NF)	w/mushrooms (R)	Soup, cream of mushroom (R)
Cauliflower w/cheese (R)	Green beans and broccoli (R)	Spaghetti w/meatless sauce (R)
Cereal, bran flakes (R)	Ham (I) (T)	Strawberries (R)
Cereal, cornflakes (R)	Jam/Jelly (T)	Tomatoes, stewed (T)
Cereal, granola (R)	Macaroni and cheese (R)	Tuna (T)
Cereal, granola w/blueberries (R)	Meatballs w/barbeque sauce (T)	Turkey and gravy (T)
Cereal, granola w/raisins (R)	Nuts, almonds (NF)	Turkey, smoked/sliced (I) (T)
Chedder cheese spread (T)	Nuts, cashews (NF)	Turkey tetrazzini (R)
Chicken a la king (T)	Nuts, peanuts (NF)	Vegetables, mixed Italian (R)

Beverages

		Condiments
Apple drink	Instant breakfast, vanilla	Barbeque sauce
Cocoa	Lemonade	Catsup
Coffee, black	Orange drink	Mustard
Coffee w/cream	Orange-grapefruit drink	Pepper
Coffee w/cream and sugar	Orange-pineapple drink	Salt
Coffee w/sugar	Strawberry drink	Hot pepper sauce
Grape drink	Tea	Mayonnaise
Grapefruit drink	Tea w/lemon and sugar	
Instant breakfast, chocolate	Tea w/sugar	
Instant breakfast, strawberry	Tropical punch	

*Abbreviations in parentheses indicate type of food T = thermostabilized, I = irradiated, IM = intermediate moisture, FD = freeze dried, R = rehydratable, and NF = natural form.

FIG. 8-4: *A list of the various foods and beverages carried aboard the NASA space shuttle Orbiter. Packaging methods are also shown.* (NASA)

Growing Food in Space

In many ways, living in a space facility is like living at Prudhoe Bay, Alaska; aboard a naval vessel; or in the various arctic and antarctic scientific research stations. Food supplies must come from elsewhere or be stored aboard. In the early decades of space living in Earth orbit, people will be totally dependent on terrestrial sources for their food supply.

The logistics of supplying space food from Earth requires the expenditure of energy in the form of rocket propellant. Weight is a prime factor in space travel. Approximately 8.3 pounds of food and water per person per day are required. However, most food contains a large percentage of water, which weighs eight pounds per gallon. Abundant water exists in space facilities because of human metabolism. Therefore, because of the economics involved

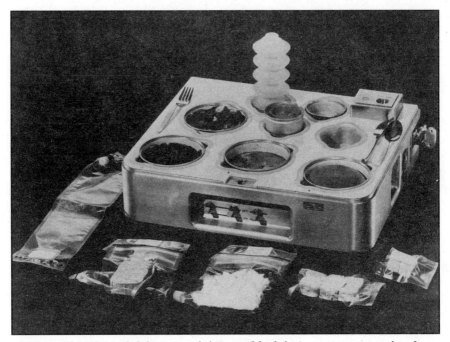

FIG. 8-5: *The NASA Skylab pioneered the sort of food that's eaten in space today: freeze-dried, foil packed, or put up in pull-top cans. The food was heated in a modular tray such as this.* (NASA photo)

(primarily the cost of rocket propellants), why lift water into space when it's already there? The first several decades of space living will see the use of dehydrated, partially dewatered, condensed, or paste foods that will then be further prepared in space using the water recovered from the life-support systems.

There will come a day, however, sometime in the first quarter of the twenty-first century, when the first meal is cooked using food grown in space.

Pioneering experiments in astroagriculture were made in the 1980s by Carl N. Hodges and his colleagues at the Environmental Research Laboratory of the University of Arizona. Plants were grown without soil on the inner surfaces of rotating drums. This simulated the environment of a rotating space station.

A 1983 study by the Boeing Company for NASA showed that growing 97 percent of the necessary food in an earth-orbiting facility occupied by as few as twelve people over a fifteen-year period could save at least $68 million in 1983 dollars.

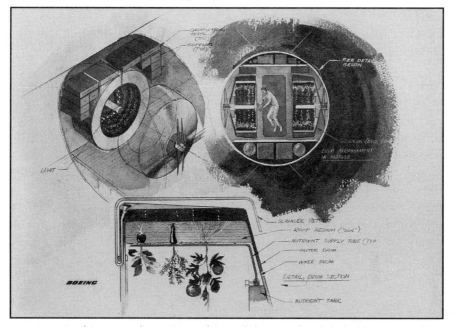

FIG. 8-6: *Food crops can be grown in the weightlessness of space. The University of Arizona and various aerospace companies such as Boeing have studied ways to do this, including growing plants in rotating drums. The first crop of weightlessness-grown wheat was harvested in the Russian space station* Mir *in 1996.* (Boeing photo)

In December 1996, U.S. astronaut John Blaha harvested the first crop of 32 dwarf wheat plants grown in the weightless environment of the Russian space station *Mir.* This was the first time that an important agricultural crop and a primary candidate for future plant-based life-support systems successfully completed an entire life cycle in the space environment. The wheat was grown in a "soil" similar to kitty litter but loaded with plant nutrients. Water was injected directly into this material. According to project scientists, no "show stoppers" were encountered that would prevent growing of other plants in larger numbers.

Food Preservatives and Preservation

However, the early space facilities will require that food be supplied from Earth. This means that fresh food is likely to be rare in space. Because of the preparation and transportation times needed to get food to space facilities, nearly everything eaten in space will be preserved food.

The art and technology of food preservation is centuries old. Meat was preserved by drying (jerky), soaking in concentrated saline solution (salting), and smoking. Spices and dehydrating have also been used.

In the twentieth century, nitrites and nitrates of sodium and potassium were widely used to cure meats and other foods because these chemicals inhibit the growth of microbes, especially *Clostridium botulinum*, found naturally in lakes and soils, which causes botulism or food poisoning. Nitrates become nitrites and are therefore used in foods that are stored for long periods. Nitrates are converted to nitrites by the chemical action of human saliva.

The use of nitrites and other food preservatives has been questioned because some studies indicated that nitrous acid created by nitrites in the human stomach's own acidic environment could cause changes in DNA that might lead to cancer and birth defects. Other studies indicated that nitrites combine with amine compounds normally found in people as well as in some foods to form nitrosamines that are carcinogens. However, many foods contain chemicals that are known to be hazardous and poisonous. For example, shrimp contain high levels of arsenic. Onions are full of highly complex carcinogen-like chemicals. Cauliflower contains a poisonous thiocyanate.

Food additives and preservatives, whether they be used for preservation or cosmetic purposes, aren't necessarily bad chemicals. A person may be forced to choose between a slight risk of cancer or a very high probability of food poisoning. Thus, the risk-benefit ratio must be considered. In spite of some scary reports about food additives, people are living longer and healthier lives eating balanced diets having a great deal of variety and choice because of preservatives and additives.

The Basic Sanitation Problem

Because humans don't use a large percentage of the food they eat and must void water produced by metabolic processes, sanitation has always been a fact of life. This is especially true when people begin to live and work together.

Some food passes through the gastrointestinal tract as roughage. The body's own waste-management system also uses the liver, kidneys, and intestinal tract to get rid of wastes and poisons that result from metabolic processes. Although a small percentage of waste is eliminated through perspiration, most of it leaves the body through urine and feces.

Any vehicle or facility, regardless of where it's located, must have some means for human waste management by a sanitary method. In other words,

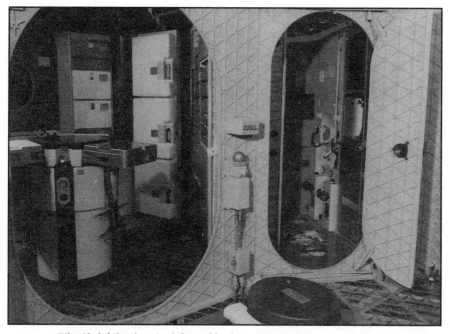

FIG. 8-7: *The Skylab kitchen (at left) and bathroom (at right) served as models for all food and waste management systems in space.* (NASA photo)

the waste must be handled so that its odors and poisons are either hidden, covered-up, contained, or processed into odorless, inoffensive, nontoxic wastes.

For automobiles, it's simple: the roadside rest stop or gas station rest room or, in an emergency, a bush.

Railroad trains simply drop everything overboard along the tracks.

Boats and ships dump everything into the surrounding water, sometimes treating or processing it beforehand.

But in the air or space, it isn't possible to pull over to the side of the road. Nor can wastes be dropped overboard. They must be collected in on-board receptacles for dumping or processing.

In early aviation—including some small, general aviation airplanes today—the provisions for handling urine were as simple as a coffee can under the seat or an overboard relief tube. But no provisions for defecation are installed in such small airplanes even today.

As airplanes got larger and carried more people on longer flights, on-board toilets were developed. The first of these was for the Douglas DC-3 in 1934. But it was only an airborne privy. The Boeing 707 jet airliner was the first to have a flush toilet; the Boeing 747 has twelve of them.

The History of Space Sanitation

Space sanitation has followed a similar development path.

Project Mercury astronauts could urinate into collection bags built into the space suits. No provisions for defecation were included, however. Astronauts were fed low-residue diets before their flights, the longest of which was a little over 34 hours. Therefore, defecation provisions didn't have to be considered in the limited weight and volume of the Mercury space capsules.

Project Gemini, however, was designed for flights up to 14 days. Astronauts couldn't be fed low-residue diets that would prevent defecation for that long. Therefore, a piece of sanitation equipment was developed that became a regular part of space travel: the Gemini Bag.

The Gemini Bag is a clear plastic bag. It contains an internal glove into which the astronaut can place his hand and attach the adhesive-ringed neck of the Bag to the anus. After defecation, the Bag is removed and immediately sealed. The Bag is then kneaded to break the seal on an inner container of germicide that, when mixed with fecal matter, prevents bacterial growth and gas formation. The Bag is then disposed of in a spacecraft cabinet or trash container.

The Gemini Bag wasn't and isn't the final answer to space sanitation, as all Gemini, Apollo, Skylab, and space shuttle astronauts learned. Astronaut Ron Evans pointed out, "There ain't no graceful way." But the Gemini Bag concept is useful for emergencies.

Skylab saw the first primitive space toilet, and the space shuttle Orbiter's toilet was a great improvement over the Skylab unit. The shuttle toilet could be used by both men and women. It could handle the disposal of urine, feces, toilet tissue, Gemini Bags, and "whoopee sacks" (or, as NASA prefers to call them, "emesis collectors"). To use it, the astronaut must be strapped to the seat to provide a seal. A flow of air replaces the force of gravity, drawing fecal matter into the depths of the commode where a motor-driven slinger shreds it and deposits it in a thin layer on the inner commode walls. The air then passes through a debris filter, a hydrophobic (water removal) filter, and fan separators. Because the space shuttle toilet must be serviced after the Orbiter returns to Earth, it's the space-going version of the jet airliner commode.

In early space capsules and facilities, urine was disposed of by dumping it overboard into space where it immediately formed a cloud of little crystals surrounding and traveling with the spacecraft in the vacuum of space.

However, the space shuttle technique of on-board collection and holding of accumulated wastes is a better way to handle the problem because a urine

FIG. 8-8: *Personal hygiene in weightlessness means keeping liquids under control and using absorbent wiping cloths. This is the wash-up station on the NASA Orbiter's mid-deck.* (NASA photo)

dump contaminates the environment around the spaceship or facility. This may affect scientific experiments, spaceship sensors, and industrial processes that depend upon a clean, uncontaminated surrounding vacuum. In the foreseeable future, waste will have to be returned to Earth.

Closed-Cycle Systems

Waste-management systems to recycle urine into potable water were demonstrated as long ago as 1959. But such "closed-cycle" systems that process wastes back into useful material won't be feasible or used in space until the economics of space living force their development and use for facilities in high orbit, on the Moon, during interplanetary flights, and on other planets and asteroids. Open-cycle systems will be used as much as possible because the technology is well-known, tested, and dependable. This is an important factor because of the high reliability required of such systems.

Experimental closed-cycle systems have been built and tested. However, maintaining ecological balance in a closed-cycle system is difficult because of

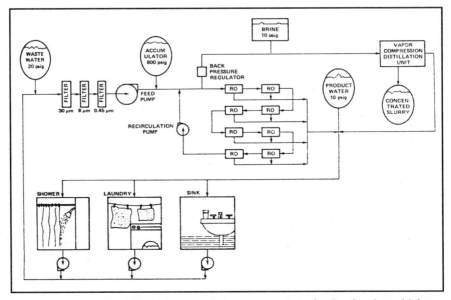

FIG. 8-9: *A generalized block diagram of the water portion of a closed ecological life sup-port system such as those required for long-duration space stays where the re-supply of consumables from Earth is expensive or difficult.* (NASA)

the small size of such systems. Earth's closed-cycle system is large and can tolerate some imbalance and repair itself if some factor grows rapidly beyond the ability of the system to handle it. In small closed-cycle systems, a minute change in a variable can cause the system to get quickly out of balance. This was one of the problems learned in the 1990s during the operation of Bio-sphere-2, a building-sized closed system located near Tucson, Arizona.

Closed-cycle environmental life-support systems will be built. Economics will drive the issue. It may have to be done in space where a massive failure cannot affect Earth's closed-cycle system. In developing such closed-cycle space systems, researchers will learn much about how to manage the Earth's system better.

But that's twenty-first century technology.

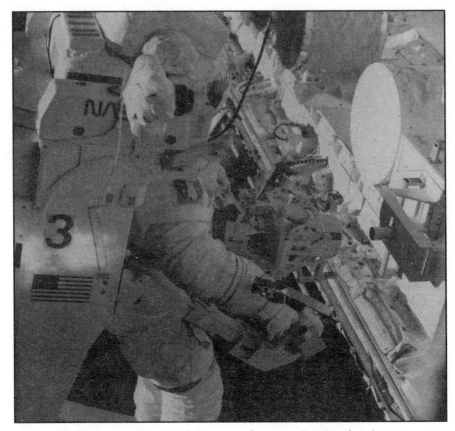

FIG. 9-1: *Working in space in weightlessness and vacuum.* (NASA photo)

CHAPTER NINE

WORKING IN SPACE

A HUMAN BEING is the best all-around tool-user ever evolved on Earth. The human body commanded by the human brain is extremely versatile. Human tool-using capabilities far outstrip those of other animals, including other primates. However, working in weightlessness presents problems because it's something human beings didn't have to do before they went into space.

The Basic Problem of Working in Space

No one should have any trouble working in space if pseudogravity is present. This is created by spinning the space facility and creating centrifugal force that the human body can't discern from gravity.

However, working in weightlessness poses some problems that astronauts and cosmonauts have encountered. To some degree, these can be recreated on Earth.

For example, with a snorkel or scuba diving gear, a person can do some experiments while under water. This is the closest simulation possible on Earth except during brief flights in high-performance aircraft.

When someone tries to replace a light bulb in an underwater swimming pool fixture while floating under water, it's quickly discovered that using a simple screwdriver to loosen the screws holding the lens is difficult. When the screwdriver is twisted counterclockwise, this causes the body to twist in the opposite direction or clockwise. Even if the diver has a nearby handhold to grasp and thus eliminates the body twisting, an entirely new set of muscles must be brought into play. Humans aren't used to using their arms to prevent twisting of the body. Therefore, some way must be found to brace the body against the twisting movement. This usually solves the problem but requires additional energy to carry out what is a simple task in gravity.

Add to this the bulky space suit that has to be worn during EVAs. An astronaut once remarked that this was like being inside a pressurized balloon, making it difficult to bend the arms and legs.

Although this phenomenon should have been anticipated for human space activities in weightlessness, it wasn't. It manifested itself on the second US space walk or EVA when astronaut Gene Cernan went outside the Gemini space capsule on June 3, 1966. The earlier EVA with Edward White on Gemini-4 had revealed no problems with an astronaut moving around in weightlessness. However, White hadn't tried to work using tools or move around the space capsule itself. Cernan discovered it was very difficult to move along the side of the Gemini spacecraft to get to its aft end. Without external handholds on the space capsule, Cernan had to rest repeatedly. The faceplate of his space suit fogged over with the additional perspiration and heavy breathing, and the suit's life-support system couldn't handle the overload. The fogging got so bad that mission commander Tom Stafford ordered Cernan to return to the capsule cockpit after about two hours of unsuccessful maneuvering trying to get to the rear of the capsule.

Wiping the fog from the interior surface of a space-suit visor is no different than having to scratch an itchy nose. It can't be done without depressurizing the space suit.

Working in weightlessness showed itself to be more strenuous than anyone had anticipated. This was confirmed by EVAs on Gemini 10 and Gemini 11 during which EVA had to be cut short because astronauts exerted themselves beyond the capabilities of their space-suit life-support systems. The problem was partially solved on Gemini 12 when Edwin "Buzz" Aldrin used body tethers, foot restraints, and handholds to brace himself while performing work in weightlessness.

Therefore, NASA embarked upon a program to learn why astronauts became over-exerted, how tools and other equipment could be properly designed for operating in weightlessness, and how to train astronauts to work in weightlessness before they went into orbit. This required more than a redesign of the Gemini space suit to make it more flexible under pressure.

Neutral Buoyancy EVA Simulation

Someone—probably a NASA engineer who had tried to replace the underwater light in a swimming pool—suggested that astronaut EVA training could be conducted under water with the astronaut in a pressurized space suit ballasted

FIG. 9-2: *NASA learned to train astronauts to work in weightlessness by using a huge swimming pool, a "neutral buoyancy simulator," that produces the closest earthbound analog to weightlessness.* (NASA photo)

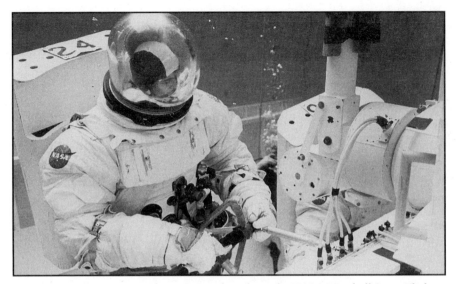

FIG. 9-3: *NASA astronaut Shannon Lucid works in the NASA Marshall Space Flight Center's neutral buoyancy water tank, learning to use a power tool in simulated weightlessness. Normal distortion makes her head look small in the transparent helmet.* (NASA photo)

to achieve "neutral buoyancy" that prevented rising to the surface or sinking to the bottom.

So the world's largest indoor swimming pool was built at NASA's George C. Marshall Space Flight Center in Huntsville, Alabama. It holds 1,300,000 gallons of water, is 75 feet in diameter, and has a depth of 40 feet. Full-scale mock-ups of spacecraft and space modules can be placed in it. Space-suited astronauts using special life-support backpacks and weighted to achieve neutral buoyancy can work in the tank and feel as if they were in weightlessness. They're monitored by scuba divers. This is the closest simulation to weightlessness that can be achieved in the persistent gravity of Earth.

Neutral buoyancy simulation worked, and other similar facilities were built at other NASA centers. Such simulators will be used for years to come to experiment with new and different ways of working in weightlessness and to train space workers.

Jet aircraft can be flown through weightless arcs in the sky, thus creating perhaps 30 to 40 seconds of weightlessness. The weightless scenes in the motion picture *Apollo-13* were filmed in a series of such 30-second weightless flights.

However, neutral buoyancy simulation permits the nearest equivalent to weightlessness that can be carried on for as long as the space suit's backpack holds out.

Working in Weightlessness

To work best in weightlessness, a person must use foot restraints. These hold the feet firmly to the "floor" as gravity does on Earth, thus permitting the normal use of hands and arms. The human body already possesses the necessary muscles and nervous system programming to overcome the reaction forces involved while using tools or otherwise working in weightlessness provided the feet are held firmly in place.

It's also possible to work well if a person can wedge the body between walls or equipment so that actions don't rotate the body. Friction still functions in weightlessness. Thus far, most astronauts have made use of this fact.

Other basic techniques applicable in weightlessness have been used by underwater free divers who manipulate tools and otherwise work on submerged equipment. Although whole body restraint is workable, most divers manage to wedge themselves between rocks or pieces of equipment to push, pull, or twist. In cases where it's possible for them to wedge their feet in place, they've discovered that they can perform most push-pull-twist tasks.

Special Tools

Situations will occur in weightlessness where it isn't possible to provide restraints to keep a person's body from moving. Although ordinary tools such as wrenches, screwdrivers, cutters, saws, and welders can be used in space, there are several zero-g tools for those jobs where ordinary ones won't work or where it isn't possible to provide body restraints.

Many of these tools are already in existence. Some of them have been used in space already; others were developed in anticipation of tasks that haven't yet been invented.

In 1965, the problems of push-pull-twist tool operation in zero-g were investigated by the Martin Company under a contract with the United States Air Force. Some of these were contemplated for use on Gemini 8 as Defense Department Experiment D-16. On this flight, Gene Cernan got into trouble during his EVA and was unable to test these tools that were located at the rear end of the Gemini spacecraft, which he never reached. Experiment D-16 was rescheduled for Gemini 11 but wasn't completed because astronaut Dick Gordon ran into problems with temperature overload in his space suit and the resulting weightlessness-induced fatigue.

Among the zero-g tools already developed are:

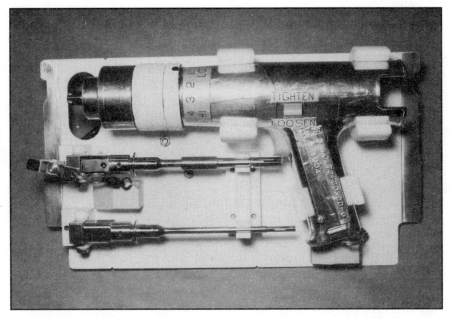

FIG. 9-4: *NASA developed several tools that could be used in weightlessness.* (NASA photo)

A zero-reaction power tool. This unit is something like the portable hand drill found in most earthside home workshops. It looks like an ordinary power drill but doesn't possess a twisting reaction force. It operates like a very fast impact wrench of the sort used by service station mechanics to remove and install lug nuts when changing a tire on a car. Designed and built by Black & Decker Manufacturing Company, this zero-reaction power tool reduced the reactive twisting torque to $^1/_{10,000}$th of its output torque through an arrangement of gearings, springs, gears, a counter-rotating barrel, and a repetition rate of 1,800 impacts per second. An ordinary service station impact wrench has a reactive force of $^1/_{50}$th of its output and operates at perhaps 10 impacts per second. In space, this would be enough to spin a space worker like a top without foot restraints. The zero-reaction power tool uses rechargeable nickel-cadmium batteries and will accept various screwdriver and socket-wrench accessories.

A ratchet screwdriver. This is like the currently available screwdriver with a ball handle and ratchet action. It was developed because an ordinary screwdriver is extremely difficult for a space-suited person to grasp whereas the ball handle is easy to hold onto while wearing a space-suit glove.

A spring-driven hammer. This hammer has a pistol grip like an ordinary rivet-driving gun except that it operates at a much higher hammering frequency.

An aluminum fingernail. This fits over the finger of a space-suit glove. It allows a person to pick up small parts. To get some idea of this problem, try to work with small parts while wearing heavy gloves. Even on Earth this metal fingernail extension is useful.

Velcro and adhesive buttons. These are placed at critical spots on the spaceship and facility so that tools equipped with compatible adhesive strips can be temporarily put out of the way when they're not being used. It's not possible to lay down a tool in weightlessness and have it stay there; it will float away. Velcro strips have been used for years aboard the NASA space shuttle Orbiter for just this reason.

Special electric lights. These may be attached to both sides of a space-suit helmet to illuminate the work area. They're also handy on Earth. Several head-mounted lights are available in hardware stores. A light with a flexible neck like the Black & Decker "Snake Light" is also useful in space.

A special chain with an internal cable. This is a hollow chain with an internal cable. When the cable is pulled tight, the chain locks into the position it had when loose. This was developed in the late 1960s at the General Electric Missile & Space Division in Valley Forge, Pennsylvania. To build one, get an ordinary set of children's play beads. At one end, attach the string securely to the end of the last bead. At the other end, install a pistol grip so that the string can be pulled taut. When the inner string is loose, the bead chain is completely flexible and can be bent and twisted at will. However, once the string is tightened, the bead chain locks itself in precisely the same position it has when the string is tightened.

A zero-g tool and parts holder. This consists of a series of small metallic or bristle brushes mounted bristles-up on a plate. Parts with holes such as nuts or small cover plates can be pushed down on the bristles to hold them in place. Tools can be wedged between the bristles.

Zero-g tool design is going to be a fun area to work in. It's totally in its infancy. Hundreds of specialized space tools are waiting to be invented to handle jobs in space that can only be imagined now.

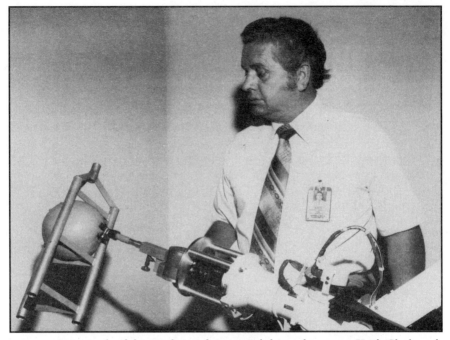

FIG. 9-5: *An example of the simplicity of some weightless tool concepts. Keith Clark used a soda straw and a toy balloon to work out the details of this simple device that can grasp and transport aluminum girders in weightlessness.* (NASA photo)

Physiological Aspects of Space Work

As discussed earlier, the human body's metabolic heat output differs from the resting state to the active working state. During sleep, the body's energy expenditure is about 65 Calories (Cal) per hour. At rest lying down, it rises to 80 Cal/hr and, when sitting up, to 100 Cal/hr. During light exercise, this jumps to as much as 200 Cal/hr and, during heavy physical work, to 500 Cal/hr. Under normal conditions, the human body is only 11 percent efficient as a heat engine. This means that 89 percent of the metabolic heat output shows up as increased perspiration rate, increased respiration rate, and slightly elevated body temperature.

Life-support systems in large space facilities are built to handle anticipated additional heat loads. But any life-support system for small space cabins and space suits must also have adequate reserve capacity to handle the times when space workers are dumping as much as 400 Cal/hr or more into the immediate environment of the cabin or suit.

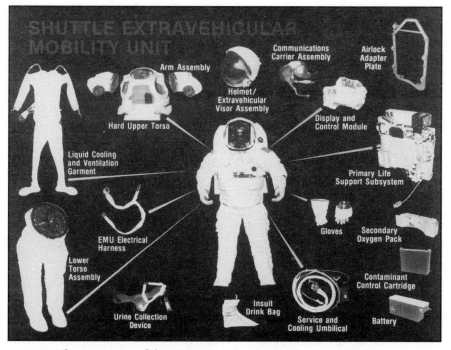

FIG. 9-6: *The major parts of the NASA space shuttle space suit.* (NASA photo)

Design of Working Space Suits

A space suit should be considered as a small, soft, body-fitting spaceship built for one. Because of its small internal volume and the limitation of bulk placed upon any self-contained life-support backpack, a space suit is extremely sensitive to the wearer's metabolic heat output as well as to external environmental stresses such as heat from sources like the Sun. The situation is less critical if the space suit gets its life-support through an umbilical from the larger system in a spaceship or facility. However, even the Gemini space suits that were attached to the spacecraft by umbilicals were limited in their abilities to handle human heat output overloads.

The space suits used in the NASA space shuttle were better in this regard. However, a shuttle astronaut still requires underwater neutral buoyancy training to learn how to pace the work and hold metabolic output levels to those that the suit can handle.

Advanced space suits may not be tailored to specific users as NASA space suits are. Like clothing, they will come in standard sizes—perhaps small, me-

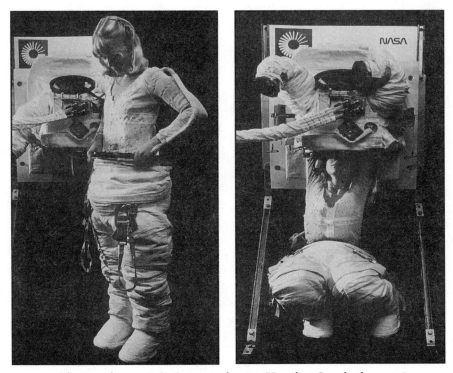

FIG. 9-7: (above and opposite) *Sequence showing Hamilton Standard test engineer Jocelyn Johnson donning the NASA space shuttle space suit.*

dium, and large. They will also have the life-support backpack capacity to handle 500 Cal/hr for at least an hour, which is the length of time an average person can work at the heat expenditure rate of 500 Cal/hr.

The life-support system is contained in a backpack because it can't be distributed around the space suit and because the backpack interferes least with the wearer's movements and working activities.

The space suit must also be easy to get into and out of like the NASA space-shuttle suit. Ease of donning a space suit is critical in an emergency.

Teleoperators

The new field of teleoperators may allow space workers to carry out some tasks without having to suit-up and go EVA.

A teleoperator is an operating machine that is directed from a remote location by a human being.

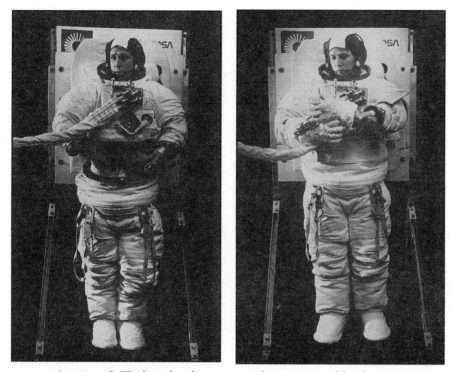

FIG. 9-7: (continued) *The liquid cooling garment, lower torso, and hard upper torso are seen in the first photo.* (Hamilton Standard)

A crude example of a teleoperated device is a radio-controlled model airplane or car. The human operator may be hundreds of yards or more from the device and yet give it commands by radio. The behavior of the machine is sensed by the human being who can then send correcting signals if necessary. A radio-controlled model airplane that carries a television camera and transmitter becomes a more sophisticated teleoperator. Radio-controlled target drone aircraft used for antiaircraft missile targets or tactical fighter gunnery training are also teleoperators.

As teleoperator technology progresses, space workers may be able to sit comfortably inside a space facility and operate a machine outside in space by a radio-controlled "up-link" to the machine and an activity-reporting "down-link" from the machine to tell the human operator how it's doing.

In a way, the early Surveyor spacecraft that landed on the Moon and scooped up samples of lunar soil were teleoperators. So was the Viking Mars lander of 1979.

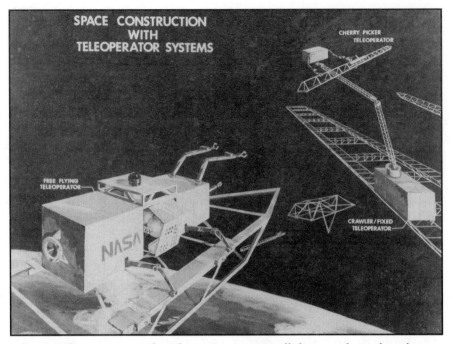

FIG. 9-8: *Teleoperators are robots that are remote-controlled on a real-time basis by people located elsewhere. They can be used in places and for work that may be hazardous to people.* (NASA photo)

A pre-programmed robotic machine is not a teleoperator. Such a device carries out its tasks according to the advanced programming stored in its on-board computer and memory. It's a robot.

Teleoperator Problems

Teleoperators haven't seen wide use because they have several basic problems.

The first of these was discussed earlier: How to design and build a machine that can do everything a human being can do. It's totally possible to build a machine that will duplicate the human arm-wrist-hand-finger actions, but it would be a clumsy device if known technology was used. Therefore, teleoperators are larger than humans and are built to do specific tasks.

Such a teleoperator is restricted in its usefulness because it will do only the job it's designed to do. If it becomes necessary for it to do something else, the human operator must either be extremely clever to manipulate it to do an unforeseen task or send in another teleoperator that's been designed for the task. The robots depicted in some science-fiction motion pictures cannot be built today.

This problem may be overcome as "nanotechnology" progresses.

"Nano" comes from the Greek word *nanos* and is a prefix that means "very small." For example, a nanosecond is one one-billionth (1×10^{-9}) of a second. Nanotechnology was forecast in a 1986 book, *Engines of Creation*, by K. Erik Drexler who was also a space advocate. He discussed and described extremely tiny machines, smaller than a human cell, that could perform useful work. Some mechanical nanomachines such as wee electric motors have been built. These are smaller than the head of a pin and can be seen only with a microscope.

In the early twenty-first century, nanotechnology may produce useful teleoperators for space work that are small, lightweight, and energy efficient.

The biggest teleoperator problem is the command circuit time delay. If the teleoperator has a radio up-link and down-link, the problem of time delay is inconsequential when the distance between the human operator and the teleoperator is within 50,000 miles. At distances greater than this, the performance of the radio command link system becomes affected by the fact that radio waves travel at the finite speed of 186,000 miles per second.

It is barely possible for a human on Earth to run a teleoperator on the Moon at a distance of about 240,000 miles. The time delay is about four seconds. This is the time required for the up-link command signal to get to the Moon and the down-link reporting or compliance signal to return to the Earth. A four-second time delay can be tolerated except in emergency situations.

Airplane test pilots know that a time delay of half a second between a control input and an aircraft response is intolerable. When it happens, it means it's time to leave the airplane by "punching out" because that airplane will soon become a smoking hole in the ground. This can be confirmed by running any of the flight-simulator programs on a computer whose CPU and monitor visual up-date speeds are too slow for the program.

Trying to run a teleoperator on Mars from the Earth is likely to be nearly impossible because a time delay of an hour or so may be involved. The teleoperator may see and drive off the edge of a martian cliff before it can be told to stop.

The Role of Artificial Intelligence

Engineers have been working on this problem since the early 1960s. The progress in computer technology has helped significantly. In a way, an electronic computer is an extension of the human mind, being able to process numbers faster and store memory for a long time—unless the computer crashes. However, computers are no better than the software (programs) that tell them what to do. It's been possible to write software that will allow a computer to be "self-programming"—i.e., to modify its program on the basis of the input. Some of this work depended on the development of "fuzzy logic" that differs from ordinary Aristotelian logic by addressing such problems as, "If A, then maybe part of B, except maybe also part of C depending on what D is." This has also had an impact on "artificial intelligence," a field still in its infancy. The computer word processing program used to write this book has a spell-checker utility that uses both fuzzy logic and a certain amount of primitive artificial intelligence to guess at several possible spellings of a word that doesn't match any in its memory; it then asks the operator to choose the correctly spelled word from among a list of possible alternative words.

Teleoperators using fuzzy logic and artificial intelligence still won't replace a human being on the spot. However, such a machine may come close. It is very useful for going where people can't, such as deep in the atmosphere of the planet Jupiter, for example.

Remote Manipulators

Simple extensions of human anatomy have been operated in space and will continue to be important in space work for decades to come.

Although extension-type tools were used by astronauts to repair Skylab in the 1970s, the Space Shuttle Remote Manipulator System (RMS) was the first to be designed for moving objects around near the space shuttle Orbiter. The RMS, designed and built in Canada, has a shoulder, elbow, wrist, and grasping "end effector." The RMS is operated by an astronaut from the Orbiter flight deck where the RMS can be seen through cabin windows. Additional views come from television cameras on the elbow and end effector.

The RMS is a space crane, the same sort of extension of human reach and muscular strength as a construction crane on Earth.

Waldos

Such extensions of human reach and strength fall in the category of devices often called "waldos" because they were first described in detail by Robert A. Heinlein in a 1940 science-fiction novel, *Waldo*. In the story, Waldo Farthingwaite Jones, who suffers from myasthenia gravis, lives in the weightlessness of an orbital facility and develops a series of power-boosted extensions of his weak fingers, hands, and arms. This was shortly before similar remotely activated arms and fingers were developed for handling hazardous radioactive materials during the development of the atomic bomb.

FIG. 9-9: *The NASA space shuttle Orbiter's "Remote Manipulator System" or RMS, built in Canada, can be used as a "cherry picker" by an astronaut. Note foot restraints.* (NASA photo)

Handling Solids and Liquids in Space

Handling solid objects in weightlessness is no real problem. A solid object retains its shape and size. However, some solids are nuisances. Electrical wire, communications cables, and other sorts of ropes and lines are messy because they float, become tangled with themselves and other objects, and get in the way. On the first space shuttle flight, astronauts John Young and Robert Crippen discovered that the wires connecting their communications headsets and units to equipment in the Orbiter cabin were getting tangled in everything, including themselves. On later flights, wireless headsets were used, eliminating the bothersome wires.

FIG. 9-10: *Man-amplifiers or "waldos" are used to increase human muscle strength or capabilities for heavy space work. They have been used in the nuclear industry since the 1940s.* (Rockwell International Corporation)

Handling liquids in weightlessness is something else because a liquid, by definition, doesn't retain its shape and can easily separate itself into smaller particles of liquid. On Earth, we're accustomed to a liquid staying in its open container provided the liquid surface is "uphill" from any opening. If the container is tipped so the opening is "downhill" from the liquid surface, the liquid therein will pour out.

Liquids don't behave that way in weightlessness.

Like a solid, a quantity of liquid will float freely wherever it's put. But, unlike a solid, it may not retain its shape. If undisturbed, a quantity of liquid will assume a semi-spherical shape. If disturbed by any outside force such as a current of air or by rotation, it becomes an oscillating glob. Disturb it to the extent that its surface tension is broken, and the glob will immediately separate into a host of smaller globs.

Because a liquid has internal forces such as surface tension, it also has a tendency to either "wet" any surface with which it comes into contact or to repel itself from any surface it can't wet. Surface tension is the reason a needle will float on the surface of a glass of water on Earth.

If a liquid can wet a surface with which it comes into contact, it will flow and cover the entire surface. On Earth, it will form a meniscus that can be seen at the surface of the water in a drinking glass; the edge of the surface bends upward where it touches the glass container. Only the force of gravity keeps the water from flowing up the surface of the glass and wetting the entire container, inside and out.

In weightlessness, water in a glass will crawl up the inside surface of the glass, over the lip, and all over the outside of the glass. It cannot be kept inside any open container it's capable of wetting.

Therefore, in weightlessness, liquids must be contained and transferred in closed vessels.

If globs of water, coffee, or body wastes get loose in a weightless cabin, they'll either break into increasingly smaller globs until the air is filled with a mist that the life-support system must deal with, or the globs will touch and wet all suitable surfaces they happen to touch.

In general, except when some laboratory monkey wastes got loose in the Orbiter cabin, this behavior hasn't caused too much of a problem in space because the surface tension of a liquid can be used to keep it under control.

The easiest way to do this is to mop up the liquid with an absorbent cloth, sponge, or paper tissue. Such materials possess large surface areas that liquids try to wet by surface tension. But once such a mop-up material is saturated with liquid, it can't be squeezed semi-dry as on Earth. It must either be discarded or used with special equipment to contain the liquid squeezed out.

During everyday living in space, moderate amounts of liquid can be handled without difficulty. It's only when dealing with large amounts such as may be used in space industrial processes that big problems may arise. But, in those special circumstances, special procedures and equipment unique to the space industrial operation hopefully will have been designed ahead of time and available. If not, space workers may run into difficulties.

FIG. 9-II: *Assembling a space facility structural truss in weightlessness.* (NASA photo)

Conclusions

Performing work in the weightlessness environment of space may be difficult and seems to pose many new problems at first. However, thus far in the history of people in space, no insoluble problems are anticipated. This doesn't mean there won't be any because, until many people are living in space for many years doing a wide variety of tasks, unforeseen problems are certain to

FIG. 9-12: *Repairing equipment in orbit.* (NASA photo)

arise. People working in space will have to learn some new ways of doing what they've taken for granted in the Earth's gravity field. But since the days of Skylab and the Soviet *Salyut* space stations in the 1970s, there is no question that human beings can perform useful work in space, although the debate goes on over the moot question about the ultimate value of people in space. Space is for people; it's a new frontier that requires new ways of doing some old and familiar things that only human beings can do.

FIG. 10-1: *Control panels and other equipment used by people in space will be carefully designed.* (NASA photo)

CHAPTER TEN

DESIGNING FOR HUMAN BEINGS

The Misdesign of Human-Operated Devices

People who developed the early, classic, and historic technologies that predate the Industrial Revolution paid little attention to the needs of the human beings who had to operate mechanical devices. At best, the design of carriages, weapons, and tools was a matter of cut-and-try, known now as "empirical development." Once the basic devices were proven to be more or less usable, the designs were not changed for centuries in most cases. Other devices based upon these original ones were strongly influenced by them and therefore perpetuated mistakes or misconceptions of human size, strength, or operating abilities.

The principle is illustrated by the steering of ancient sailing vessels. This was done by a helmsman operating an oar that served as a steering rudder and, later, the hinged sternpost rudder. These steering mechanisms were located at the aft end of the hull. So was the helmsman. The aft end of a vessel isn't the best place to put the person who's responsible for seeing where the ship is going and steering it safely through the rocks and shoals. Even when blocks, tackle, and other mechanisms were introduced, the helmsman remained at the rear of the ship and wasn't moved to the front because the tradition had been firmly established. "If it works, don't change it."

It wasn't until the new technology of steam power demanded a complete overhaul of ship design that the helmsman was repositioned to a bridge on or above the forecastle where he could see where the ship was going.

Imagine trying to drive a bus from the rear.

This is only one example of the misdesign of devices for human beings.

Difficulties are always faced by farsighted advocates who try to change the basic design of a human-operated device to make it more sensible. Steam railway locomotives were operated with the engine driver located in a cab at the rear of a boiler that blocked most forward vision. Although this was primarily because of the need for communication between the engineman and the fireman, much of this need for communication was caused by the fact that the engineman couldn't see very well and had to depend on another set of eyes on the other side of the locomotive. Some cab-forward steam locomotives were used on the Southern Pacific railroad from 1928 until the 1950s, but these were ordinary steam locomotives turned end-for-end because they used liquid oil that could be piped to the fire in the boiler, not solid coal that had to be shoveled or transferred by a feed screw from the tender. It was only when the Diesel-electric locomotives came along that a major technology shift resulted in the engineman and fireman both being positioned at the front of the locomotive. Even then, the crew didn't feel comfortable without part of the locomotive in front of them. The result can be seen today in the modern Diesel-electric locomotive such as the General Electric U34 type with its cab located partway back from the front end.

Many machines were designed on the basis of tradition as well as ease of construction.

Why does the automobile have its engine in front? Because that's where the horse was located that pulled the surrey, buckboard, brougham, or coach. Designers placed the engine in front of the car to drive the rear wheels through a drive shaft arrangement because it was easier to build that configuration than to work the bugs out of the complex flexible joints to transmit power to the front wheels that also had to be turned for steering. A partial solution came with the relocation of the engine to the rear of the car, and several very fine automobiles were produced with rear engines. Although the Chevrolet Corvair was a perfectly good car, it came afoul of an insignificant oversight in design that could easily have been corrected. However, this was not done because of political and emotional pressures. When the technology finally arrived at the point where it was possible and economical to build practical front-wheel drive autos, almost every manufacturer began to produce such cars.

The Empirical Design of Clothing

In the past, machines weren't the only items not properly designed for human beings. The very clothing that people wore to keep warm in cold climates, keep cool in warm locales, and provide a surrogate skin as an antichafing layer was a matter of empirical construction.

Although some clothing such as that developed by the Eskimos and other arctic peoples was and is well-fitted and extremely efficient, it's typical of clothing everywhere that it must be individually tailored to the individual human being to fit well. In the past, peasants, slaves, and the impoverished lower classes wore clothing that was not much more than ill-fitting rags because tailoring was and still is expensive. Clothing was worn until it literally wore out. Only in tropical and semitropical climes was clothing washed. Most Europeans learned from people in the Orient and tropics to wash clothing and themselves to promote personal and public hygiene.

People in the past are often considered as being well-dressed. But the image comes from highly idealized portraits and paintings of people rich enough to have their clothing made individually for them by tailors. A close look at clothing from the past preserved in such museums as the Smithsonian Institution reveal that it's ill-made, ill-fitting, and extremely coarse by modern standards.

Anthropometry and Ergonometrics

Why were these shortcomings tolerated? Why were people satisfied with machines that were poorly designed and difficult to operate? Why did they put up with ill-fitting clothes?

The answer is simple: No one knew how big people really were, how the sizes of their limbs and torsos varied, how far they could reach, how they sensed their surroundings, how long it took for them to see something and act, how their vision was limited, and how long they could maintain top efficiency while doing some physical activity. No scientist or technician had measured human beings.

It was only in the twentieth century that the modern science of *anthropometry* began. This is literally the science of people measurement.

An offshoot of anthropometrics is a field of technology that has only recently been given the name *ergonometrics* or "work measurement."

People Measurement

Anthropometry had its beginnings in the military services.

In comparison to today, armies of the past were relatively small before the First World War. Napoleon's *Grande Armée*, probably the largest military force assembled in history to that date, never numbered more than 500,000 men at any given time. The battles of the Napoleonic Wars in Europe rarely saw more than 150,000 men in action on each side. This was partly due to the enormous difficulty of communicating with very large military forces spread over an extensive piece of real estate. A logistics problem existed as well in supplying these men with ammunition and meager amounts of food. Most armies had to live off the land to eat.

Someone had to make all the uniforms for these soldiers. In those days, it was the work of tailors. Not all uniforms fit as well as contemporary artists portrayed them. One of the purposes of a uniform was to visually identify friendly and enemy troops. Most soldiers had only one uniform to wear. Officers usually had their uniforms tailor-fitted while the troops were kitted-out in uniforms that might not fit as well but served the identification function on the battlefield.

As the population increased in the nineteenth and early twentieth centuries, mass production of clothing became necessary for civilians as well as soldiers. Most of it was incredibly ill-fitting. Old photographs confirm this. The 1900 Sears Roebuck catalog featured mail-order clothing that was made to order from fifteen measurements of each individual that had to be sent with the order.

When millions of men were called to arms in Europe in 1914, it became necessary to make uniforms for these soldiers. The task would have swamped all available tailors. The requirement gave birth to the mass-produced clothing industry. Most of the uniforms were made to fit what army officers considered "average men." But no one knew what size the "average man" was and had no data upon which to base the sizes of the uniforms that were ordered. When "small, medium, and large" was introduced, no one knew how many of each size to order. As a result, most of the uniforms didn't fit or they were "loose fitting" on the basis of "one size fits all." If a soldier was lucky, he either had a little pocket money to get his uniform fitted by tailors in the nearby town or had to use his first military pay for that purpose. (Privates in the US Army got nineteen dollars a month then.)

The experience of the US Army with poorly-designed, ill-fitting, and otherwise ineffective uniforms during the First World War was sobering. Far too many soldiers froze to death, suffered from frostbite, or got trench foot in France because of the poor military clothing and equipment.

But for the first time in history the army had gathered physical data on its inducted men. The data couldn't be used in the First World War because there was too little time to put it to use in the face of war requirements. In the 1920s and 1930s, the existence of this data led slowly to the establishment of anthropometry. In spite of the jokes about uniforms that didn't fit (most of them retreaded from the First World War), the uniforms of the soldiers and sailors in the Second World War were better fitting and more effective in protecting them against the rugged outdoor environment.

To a person from 1900, the uniforms of today's soldier, sailor, or airman would be considered outstanding examples of tailoring. As for the fantastic clothing and equipment available to the modern camper and outdoors person, Napoleon or Ulysses S. Grant would not have believed the contents of the L.L. Bean catalog.

The reason clothes fit so well today even though they may have been bought off the rack in a big chain department store is that now clothing manufacturers know how big people are and how many need small, medium, large, or extra large.

The Size of Human Beings

The drawings in Figure 10-2 illustrate how thorough this anthropometric data has become. Dimensions shown are those for a "95th percentile man." This means that 95 percent of the current adult male population has physical body dimensions equal to or less than those shown.

The amount of data is now so large that there are similar data bases for women as well as for children of various ages. In addition, the variation or statistical spread of the data is well known so it becomes possible to say that so many million male adults are in each percentile of the overall adult male population.

For the first time in history, the size of human beings—at least in the United States—is known.

This data changes, and all the reasons for the changes aren't well known. People are getting larger. Whether this is due to better nutrition or to genetic changes is unknown. A medieval Englishman, based on the sizes of the armor that has survived, was about 5 feet 7∫ inches tall. In World War II, this had increased to 5 feet 8° inches. Today's 95th percentile male army soldier is 5 feet 9 inches tall. If colonial houses in New England seem diminutive, it's because people today are bigger than their ancestors.

Using Anthropometric Data

The data are used to design more than clothing although proper clothing is important to space workers.

The NASA space shuttle space suit (NASA calls it the "extravehicular maneuvering unit") was designed making extensive use of anthropometric data.

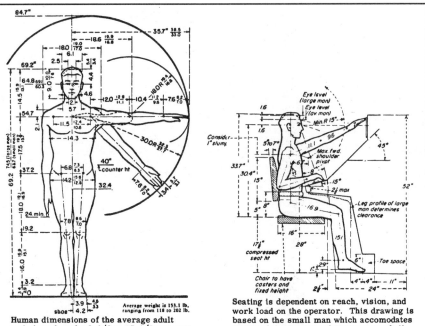

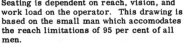

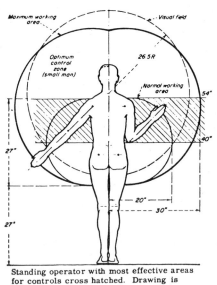

Human dimensions of the average adult male based on the 2-1/2 to 97-1/2 per cent range of measured subjects.

Seating is dependent on reach, vision, and work load on the operator. This drawing is based on the small man which accomodates the reach limitations of 95 per cent of all men.

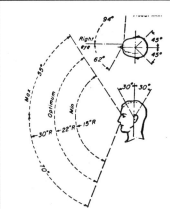

Visibility studies must include considerations of size, color, lighting, and purpose of the equipment. Where concentrated attention will be required, effective visual areas are considerably reduced.

Standing operator with most effective areas for controls cross hatched. Drawing is based on small man (2-1/2 percentile).

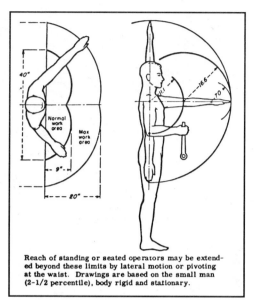

Reach of standing or seated operators may be extended beyond these limits by lateral motion or pivoting
at the waist. Drawings are based on the small man
(2-1/2 percentile), body rigid and stationary.

FIG. IO-2:

Human dimensions. (US Air Force)

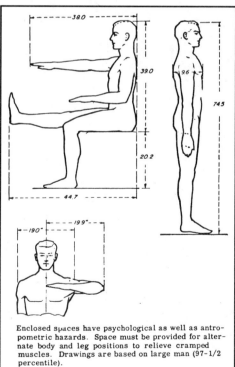

Enclosed spaces have psychological as well as antro-
pometric hazards. Space must be provided for alter-
nate body and leg positions to relieve cramped
muscles. Drawings are based on large man (97-1/2
percentile).

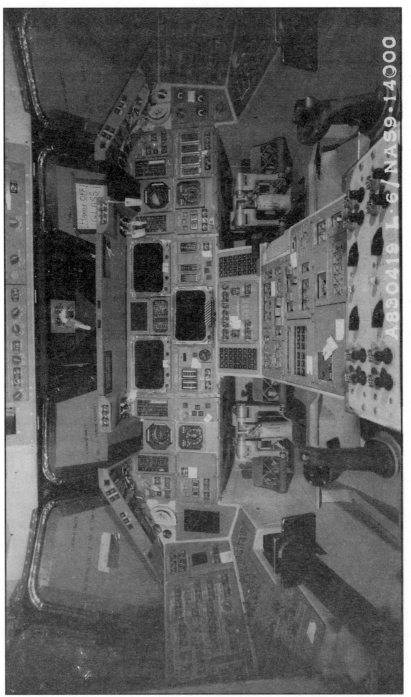

FIG. 10-3: *The cockpit and instrument panel of the NASA space shuttle Orbiter look complex but were designed with human factors in mind.* (NASA photo)

The lower torso or pantslike portion comes in a variety of different sizes and attaches to the hard fiberglass upper torso by means of a beltlike metal waist ring. The upper torso comes in five different sizes. The gloves are available in fifteen different sizes, but the bubblelike helmet comes in only one size. The various parts of the suit were designed with the "small/medium/large/extra-large" philosophy while the helmet that doesn't have to fit tightly was designed for the 97.5th percentile—i.e., the helmet will fit 97.5 percent of the population.

A space suit is one item of space equipment that will continue to depend on anthropometric data.

Anthropometric data was also used to design and lay out the flight crew stations. For example, complex as it may appear, the instrument panels in front of the commander/pilot and copilot were carefully laid out with anthropometric data. So were the various control panels at the rear of the flight deck where the payload bay experiments and RMS are operated.

The use of anthropometric data in the design of vehicle operating positions is quite new. Propeller-driven airliners had complicated instrument panels and the location of each dial and switch was a matter of "stick it where it fits." This was also true of small general aviation airplanes until as recently as 1970.

Principles of Ergonometrics

Human beings who operate devices must be considered as both *sensors* and *responders* for the machine as well as the cognitive *directors*.

Treating a human being as a machine doesn't necessarily mean dehumanizing the individual. In fact, considering a human being as a part of the overall system makes it possible to design it so its operation becomes a more humane activity.

In spite of a person's limited sensing range to various external stimuli, on the basis of size and weight a human has far more sensors of far wider overall range than any man-made device. Furthermore, a human can exercise cognitive direction without pre-programming. This is the unique ability of a person to come up with answers to problems that were unforseen and unanticipated.

As Robert A. Heinlein once pointed out in a private conversation, it would be extremely difficult to design a robot truck driver capable of leaving the Long Beach docks in California and, several days later, passing through New York City's Holland Tunnel with an accuracy of a few inches, having in the meantime coped with millions of unanticipated decisions en route. Some military cruise missiles and "smart bombs" could come close. However, in the meantime the machines of space must be designed for people.

Human Reaction Delay Times

In designing machines, especially space vehicles where high relative speeds may be involved, the time delay inherent in the human nervous system must be taken into account.

The time required for a person to *perceive* the reception of a stimulus depends upon the intensity, duration, contrast, and relationship to unwanted signals that can be termed "noise." It also depends upon which sensory organ is stimulated. Threshold values below which a stimulus will not be perceived are important. A low-intensity stimulus may not be noticed. If the time duration of the stimulus is too short, it won't be perceived. People don't notice the dark screen that occurs between the sequential projection of frames of a motion picture nor the 60 hertz flicker of a fluorescent lamp. If the signal has an intensity very close to that of the background, it can't be discriminated from that background; this is one of the principles of natural and military camouflage. If a high level of noise is present, the stimulus may be buried in the noise.

As intensity, duration, contrast, and signal-to-noise ratio increase, the time required for a person to sense a stimulus decreases up to the point where the particular sense organ is operating at maximum efficiency. Beyond that point, there is no decrease in perception time. At high levels, the sense organ may become overloaded. In the case of duration, the stimulus may occur over such a long period of time that the human system doesn't recognize it as change because it takes so long to happen.

The major elements that human senses detect are *change* and *rate of change*. This holds true for other mammals as well.

Time delay within the nervous system itself is variable, depending upon the type of stimulus and how it's perceived.

The most important sense organ for space activities is the eye. It requires 20 milliseconds (0.020 seconds) to respond to a visual signal under optimum conditions of light intensity, duration, and contrast. The transmittal of this information through the retina and optic nerve to the brain requires an additional 2 milliseconds. Once the nerve signal reaches the brain, the cortex requires about 13 milliseconds for excitation. This total of 35 milliseconds is required for the *perception* of a visual signal. Under conditions less than ideal or if the person is distracted by other stimuli or activities, this sensory time delay for visual stimuli can be as long as 100 milliseconds or a tenth of a second. Furthermore, the nervous system doesn't seem able to handle a rate of data input greater than 10 bits per second, which is why people don't see the flicker of a motion picture projected at 24 frames per second, the successive

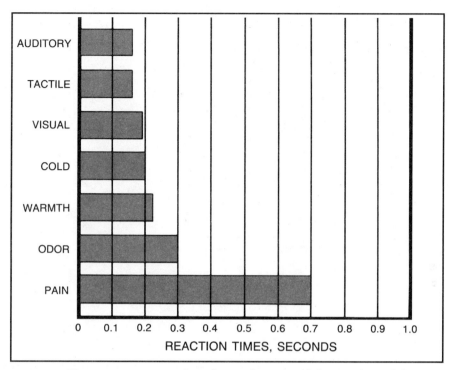

FIG. 10-4: *Human reaction times depend upon the nature of the stimulus and the human sensor involved.* (Art by Sternbach)

projections of a television raster at 30 frames per second, or the 60 hertz flicker of a light.

In the simple case used as an example above, however, only a single stimulus was assumed. Because of the limiting rate of information transfer of the visual system, it can be overloaded easily by overlapping, competing, or incomplete signals. This leads to confusion.

Responding to a stimulus requires more time, depending upon a person's physical condition, psychological set, level of alertness, motivation, and emotional state. Reaction time is also dependent upon the nature of the response—forcible movement and displacement versus precision manipulation. Since most human reactions are motor responses such as pushing buttons, turning knobs, moving switches, and activating levers or pedals, the limb being used to respond also affects the reaction time. Eye-hand actions for simple tasks are about 20 percent faster than eye-foot actions. For a right-handed person, the eye-hand reaction time is about 3 percent faster for the right hand than for the left hand.

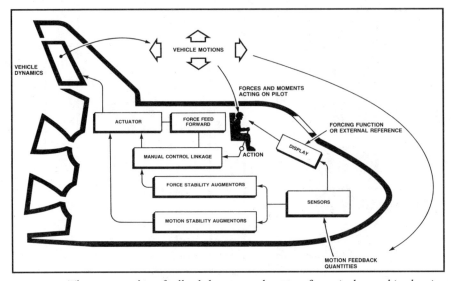

FIG. 10-5: *The man-machine feedback-loop control system of a typical spaceship showing the various feedback loops.* (Art by Sternbach)

The reaction times shown in Figure 10-4 are those for simply pushing a button in response to various types of stimuli. The usual stimuli are auditory and visual. These have total reaction times of about 160 milliseconds and 190 milliseconds respectively. Response to an odor or to pain takes longer.

The response to a more complex task such as piloting a spaceship during high-rate orbital closure or landing is considerably longer. When added to the total response time of the nonhuman system being controlled, the total lag time for the human-machine system can be as long as ten seconds.

Taking time delays into account, two spaceships each having a ten-foot diameter closing on one another at 3,000 miles per hour will collide before either pilot can see the other object, recognize it, evaluate the situation, decide what to do, and take evasive action.

Therefore, computer systems that are much faster are used to supplement human perception and reaction in high performance military airplanes as well as spaceships.

Human-Machine Systems

If human shortcomings are to be overcome and the optimum use of both the human and machine achieved, designing the machine around the human being is therefore an absolute necessity not only for space devices but also for other technologically advanced and highly complex systems.

As the discussion about teleoperators pointed out, the human operator must have timely information about what the machine is doing and how it's responding to commands. In technical language, this is termed "feedback." It's the principle on which most human-machine systems operate.

A simple feedback system is shown in Figure 10-5. When the human sends a command to the machine, the machine reports back that it has received the order and then carries it out. It then reports back to the human that the task has been initiated, that the machine is responding, and that the commanded task has been completed. If the machine encounters problems in doing any of these, it reports immediately to the human so that other remedial actions can be commanded. As discussed, if the time delay in the feedback loop is too long, the machine may be doing something different by the time the human can send a correction signal to it. Such long-delay systems are said to "go hyperbolic," which means they get totally out of control and exhibit increasingly severe departures from the desired activity. This is what causes a test pilot to bail out of an experimental airplane when the time delay exceeds half a second.

Because of the millisecond delay times in the human nervous system, many machine systems now include a control computer that operates in parallel with the human operator. The human can command the computer or override any computer command, but the computer does the repetitive jobs that keep the machine running properly. The human operator then commands only *changes* either to the computer or, by overriding the computer, directly to the machine.

The automobile cruise control and the airplane autopilot are earthly examples of this sort of computer-assisted control. The cruise control is a computer that senses car speed, compares this datum against the predetermined cruise speed programmed into it by the driver, then operates the engine throttle to correct any speed error. Airplane autopilots come in various models from those that will simply keep the wings level to ones that will level the wings, hold a compass heading, track a radio navigation signal, follow a programmed course using satellite positioning signals, and maintain altitude. The most advanced airplane autopilots use inertial guidance and satellite positioning; they can fly an airplane from takeoff brake release to landing rollout—but always with human operators in the command override positions.

This has led to a new field of science and technology called "semiotic analysis." It makes use of such mathematical advances as fractals and chaos theory in the design of control mechanisms for complex systems. This work has applications in social and political systems as well.

Principles of Display

A person's ability to operate in a man-machine system, regardless of the degree of automation and computerization involved in the system, depends greatly on what information is displayed to the operator and how it's presented. This, too, is part of anthropometry and ergonometrics because it involves how a person perceives things and what sort of information displays provide the best information under a variety of circumstances.

Spaceships are more analogous to airplanes than automobiles, but many of the same display principles apply to all three types of transportation vehicles.

Automobiles in the 1950–1970 period had excellent displays of information necessary for a person to properly operate the machine. These included an analog or moving pointer speedometer, a digital odometer, and analog dials presenting data on engine coolant temperature, engine oil pressure, battery-generator current, and fuel supply. Automobile design then went through a period when most displays deteriorated to digitized speed presentations plus "failure indicators," more commonly called "idiot lights," that were capable of reporting that engine temperature, engine oil pressure, electrical system, or the brakes had exceeded predetermined limits or failed. No means were provided to indicate an imminent failure because of, say, rapid increase in coolant temperature. When the failure happened, a person could only hope that the car could be driven to the side of the road without unduly endangering the driver or passengers.

Airplanes are different. If something quits, it's not possible to pull over to the side of the road. Therefore, federal law requires that critical airplane and engine performance and condition be monitored and displayed before the pilot. In some of the more recent designs, an instrument dial pops up on the television-like panel viewscreen only when the measurement item exceeds a specified value or begins to change rapidly. Idiot lights are the bane of an airplane pilot's life, especially in airplanes with retractable landing gear where the position of the gear, invisible from the cockpit, is reported by the opening or closing of position switches on the gear and displayed in the form of lights on the instrument panel. Gallons and gallons of cold sweat have been shed because the light bulbs were burned out, the switches jammed with mud or frozen with ice, or the wiring connections come loose. There is no airport in the country that hasn't experienced a pilot making a "low pass" at low altitude down the runway so that another pair of eyes on the ground could check the landing gear position.

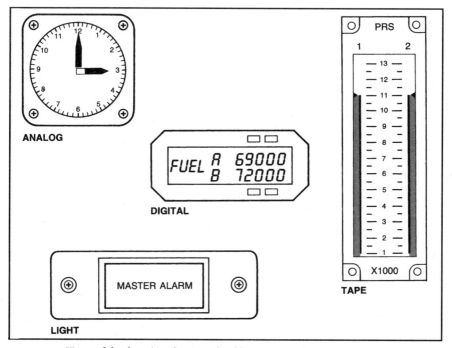

FIG. 10-6: *Types of displays.* (Art by Sternbach)

Types of Displays

Analog, digital, and binary are the three types of information displays.

An analog display is usually a circular or semicircular gauge with a moving pointer such as the one shown in Figure 10-6. So is a type of gauge called a tape that is a circular gauge straightened out as also shown in Figure 10-6. A clock with moving hands is an analog time display.

A digital display is a presentation of a row of numbers. The mechanical odometer on an automobile instrument panel is a digital display of distance traveled since the driver reset it. A pocket calculator has a digital display. Digital clocks are commonplace.

A binary display is an idiot light that's either on or off.

Each type has its advantages and disadvantages.

An analog display isn't accurate; it's an indication. It may be possible to read it to three significant numbers. However, greater accuracy than this may not be required. The important datum shown by an analog display is the *relationship* of the displayed data to some maximum, minimum, or average value—the "red line" maximum or minimum or "in the green" for the average or desired value. Most important, an analog display will indicate *rate of change* by the speed with which the needle or dial indicator moves across the face of the instrument. It's often important to know rate of change. For example, in a car going up a long hill on a hot summer day, an idiot light will inform the driver only that the engine has gotten too hot. A temperature gauge permits the driver to monitor the rate at which the temperature rises. If it goes up slowly, it may be possible to reach the top of the hill before the engine over-heats. However, if it goes up rapidly the driver knows that a stop will be necessary partway up the hill to allow the engine to cool.

A digital display is as accurate as its sensor system and will report to as many significant figures as it has available on its display readout. It's useful where high precision is required. But red-line values must be remembered by the driver or automatically recalled by a computer. A digital display can be useless if the data changes so rapidly that the digits become a blur. An example is a digital stopwatch accurate to $1/100$th of a second; when it's running, the rapidly changing right-hand digit can't be read.

The advantages and disadvantages of the binary display have been discussed. Any binary on-off display requires a test mode so the human operator can be sure the light bulbs are working. A binary display system should have redun-dant sensors and data transmission means in case the display lights are okay but some part of the rest of the system fails.

Redundancy and Triangles of Agreement

The need for redundancy or "back up" leads to the subject of "triangles of agreement." This is based on the ancient wisdom, "If you desire to give a friend a clock, give him three so that he will know the hour." This is not the ordinary principle of triple redundancy—"What I tell you three times is true." With one clock a person has only a general idea what the time is. With two clocks, they will rarely be in agreement and the correct one may not be discov-ered until it's too late. With three clocks, one can be reasonably sure that the two in closest agreement with one another are probably right.

Every critical display should operate on the triangles of agreement principle.

Triple redundancy requires that there be agreement among three separate systems before the data can be trusted. If one system doesn't agree with the other two, something's wrong. Triple redundancy also can mean the operation or activity is a no-go situation without agreement between the three systems. Very critical operations often involve quadruple redundancy or higher so that three of the instrumentation systems will be in agreement, an extension of the triangles of agreement principle.

FIG. 10-7: *Modern airplane cockpits make extensive use of human factors in the design of displays. This is a typical "glass cockpit" for a business jet.* (Sperry Flight Systems)

These two principles can mean that the displays for complex devices such as nuclear power plants, communications switching networks, large networked computer systems, automated factories, airplanes, spaceships, and space facilities can become very large, very complicated, and thus practically useless to a human being because of sensory overload. Such "wall-to-wall data" may be considered necessary, but it can't be assimilated. This has led to new techniques of data presentation and display in the growing field of semiotic analysis.

Displays for Complex Devices

When a person is operating even a simple machine such as an automobile, it's not necessary that all the speed, distance, fuel, and engine operating data be constantly displayed. When the data falls outside a predetermined value or rate of change, it must then be presented in a vivid, attention-getting format. This data should also be available any time the operator wants to see it. Or when the computer system that's tracking all this data fails in any way.

Displays of this sort are now commonplace on the newer jet airliners, in military aircraft, and on the space shuttle Orbiter flight deck. Information is displayed on a screen where the operator can select what's shown. Or the computer will display any parameter that wanders out of its predetermined limits. The computer can also analyze the situation and recommend actions to be taken by the human being.

Control Mechanisms

Controlling or overriding the control computer has historically been accomplished by a human operator pushing or pulling a lever, positioning a control stick, pushing a button, turning a shaft or knob, operating foot pedals, or a combination of these actions.

Two new control mechanisms have appeared in the past several decades.

In army attack helicopters, the guns can be aimed by the pilot moving his head and looking at the target. The helmet senses these movements and the direction of sight.

Another control mechanism involves the operator entering commands into a control computer. These commands are either action commands, activation of a new computer program stored in memory, or new data for the computer to use.

Summary

Of course, all controls and displays must be located and arranged on the basis of the foundation data of anthropometrics and ergonometrics. No matter how complicated the machines of space become, they can be used and controlled by human beings because we now know the most efficient ways for people to perceive and respond to stimuli.

FIG. II-I: *Fun in zero-g.* (NASA photo)

CHAPTER ELEVEN

RECREATION IN SPACE

Although people will control space equipment using computers as essential backups, assistants, and extensions to human capabilities, an important difference exists between the capabilities of human beings and computers.

Computers can operate constantly. They don't need to stop for food, rest, or recreation. Except for an occasional period of downtime for repairs, computers work three shifts around the clock without problems.

In spite of all the anthropometric and ergonometric techniques, people cannot continue to operate efficiently or effectively without rest and recreation although they're given time away from their jobs to eat, sleep, and take care of bodily needs.

In some of the early space flights, astronauts were given too much work to do and not enough time off. This caused problems, especially with one of the Skylab crews who rearranged their work schedule to suit themselves and thereafter performed even better, much to the chagrin of the earthbound people in mission control who didn't want the astronauts to deviate from the schedule at all.

"All work and no play makes Jack . . . and ulcers." That's the modern paraphrase of the ancient paradigm.

The Need for Recreation

The armed services have known the importance of recreation for years, especially the US Navy where officers and sailors must perform at high efficiency while on watch aboard seagoing vessels. The former Strategic Air Command of the United States Air Force also had to ensure that people could operate at top physical and mental condition during long watches in missile silos and on long-range bomber missions. (The US Army isn't faced with quite the same problems because they don't have to worry about having a battlefield sink or crash.)

This has a strong bearing on how people perform under the stress of living and working in space. Although many of the results of decades of research into human work efficiency may seem self-evident, the USAF Wright Air Development Center's Aero Medical Laboratory, the Holloman AFB Aero Medical Field Laboratory, and the Randolph AFB School of Aviation Medicine have come to the following conclusions regarding human performance during long and intense tasks:

1. When doing a known task, a person's efficiency starts high and then drops rapidly to a plateau of lesser performance where it remains for the remainder of the task.
2. The more intense, bright, large, or loud a signal appears to be, the easier it is to detect and respond to; there is little deterioration of performance in this regard over long periods of time.
3. The longer the watch period, the more likely that signals will be missed toward the end of the watch period.
4. Although there may be marked differences in performance between individuals, the ability to stay alert over a long period of time doesn't seem to be associated with skills or personality traits, although better watchstanding performance seems to be related to an introverted personality.

Experimental Verification

The results of several USAF experiments confirm these conclusions.

One experiment measured the performance of several human subjects doing a complex task during a continuous twenty-four-hour test. Performance peaked during the first twelve hours then suffered a continuous deterioration during the last half of the test.

Experiments were made involving individuals asked to perform thirty hours of continuous simple mental activity, allowing a five-minute rest every hour and twenty-minute eating periods every six hours. The results showed the same pattern of performance. After about sixteen hours, performance dropped from the high plateau of the initial test period, reached a low point twenty-two hours into the test, recovered to a brief but lower peak at twenty-six hours, and was again deteriorating at the end of the thirty-hour test.

Fatigue

Although some of the deteriorated performance exhibited during these tests could be attributed to boredom (lack of motivation) with the repetitive task, most of the deterioration was probably caused by accumulative fatigue, which may be a more important factor than lack of motivation.

The 1927 New-York-to-Paris flight of Charles A. Lindbergh is an excellent example in the real world. Lindbergh piloted a marginally stable airplane under stress for more than thirty-three hours without rest after having had little or no sleep during the twenty-four hours before the flight. On the basis of experimental data obtained since then, one could certainly speculate on how much longer Lindbergh could have flown *The Spirit of St. Louis.* He landed at Le Bourget Aérodrome outside Paris with enough fuel remaining aboard to fly for an additional ten hours. Whether the airplane could have outlasted the pilot is a question that will never be answered. However, the flight was a triumph of human endurance and shows what a person can do when properly motivated.

However, accumulative fatigue produces some strange side effects, including hallucinations. The USAF School of Aviation Medicine tests produced some fascinating results. In a thirty-hour extended-task experiment with subjects in an airplane flight simulator on the ground, the following "perceptual aberrations" were reported by the individuals in their subjective verbal reports:

Occasionally the RPM indicator seemed to have a little man showing head and shoulders, in a sombrero, holding an umbrella overhead.

At one time, the instrument panel faded out as such and became an olive-green fabric with a coarse sackcloth weave, solid in color, with no instruments.

The needle of an instrument appeared to stand still and the face of the instrument and the whole panel seemed to revolve around the needle.

The lower three dials seemed to be decorated as store windows—three-sided backgrounds of billowy, multicolored material with little dolls or puppets placed in each.

Similar hallucinations were reported by other scientists researching the effects of extreme fatigue, sleep deprivation, sensory deprivation, and hallucinatory drugs.

Drugs themselves don't offer any help in performing such long, fatiguing tasks. Dexedrine and caffeine produce striking initial results that improve human performance for about two hours. But, over the long run, this high initial peak of performance created by stimulants rapidly deteriorates until at the end of the task the person still performs at a significantly lower level.

It's been experimentally confirmed that a good night's sleep will produce a complete recovery of a person's ability to perform at a high level of mental intensity. In addition, a person starting work on a task soon after such a rest performs better.

The Circadian Rhythm

Another factor that enters into a person's ability to perform complex tasks involving decision making and judgment is the natural cycles of Planet Earth including the day-night cycle. This affects the *circadian rhythm* or the body's inner biological scheduling.

People are basically creatures of habit whose activities are determined by the local time of day or the position of the Sun in the sky. Many bodily functions change with the circadian rhythm—body temperature, blood pressure, stomach and intestinal peristalsis or muscle activity, kidney and liver functions, blood chemistry, and the condition of the nervous system, for example.

Biological rhythms are known to exist in practically all living creatures on Earth, including plants. Several natural cycles may play a role in determining biological rhythms—the 24-hour solar rhythm, the 24-hour 50-minute lunar tidal rhythm, and the 23-hour 56-minute sidereal rhythm being perhaps the primary ones. Unknown effects may come from the 14-day semimonthly, the 29.5-day synodic, and the 365.25-day annual cycles.

Recent research findings, some of them still controversial, indicate that the basic inner human biological clock—equivalent to the internal timer or clock of a computer—may be paced by the 8.73 hertz cycle of the Earth's magnetic field. High in the Earth's upper atmosphere is a layer of charged particles (not the Van Allen belts) called the ionosphere. It acts like a mirror to certain high frequency electromagnetic signals. Basically, the Earth is a huge electromagnetic resonance chamber. Its resonant frequency is 8.73 hertz, called the Schuman Resonance after the European scientist who discovered it. This frequency may vary from about 5 to around 15 hertz, depending upon what solar particles are doing to the ionosphere. Dr. Siegnot Lang of Germany discovered that the Schuman Resonance is sensed by the human nervous system because the Earth's magnetic field is nine orders of magnitude stronger than a nerve impulse. The Schuman Resonance provides the basic timing signal that sets the human body's internal clock, which is why Lang called it the "Zeitgeber" or "time giver."

Jet Lag or Circadian Asynchronization

The disruption of a person's circadian rhythm caused by moving rapidly from one terrestrial time zone to another is known as "jet lag." This is a new factor in human living because moving rapidly from one time zone to another wasn't possible until the twentieth century. The three-hour time difference caused by a transcontinental flight normally doesn't throw a person's circadian rhythm too far out of synchronization with local time. However, a six-hour transatlantic time zone shift does.

The physical effects of circadian asynchronization aren't as pronounced as those of fatigue although the symptoms bear some resemblance. A person's body is disrupted and out of phase with what everyone else is doing. When he or she is ready for dinner, others may be ready to go to bed.

Circadian asychronization can result in a deterioration of a person's ability to make good decisions as well as affecting judgment.

An eight- to twelve-hour rest and sleep period permits the body to achieve circadian resynchronization or "autophasing."

Circadian asynchronization among airline transport crews is prevented by having the crews remain on the time zone of their flight's origin even during the twenty-four-hour layovers for, say, a transatlantic flight.

Daily Cycles in Space

To date, people in space haven't been affected by circadian asynchronization because, like airliner crews, they've continued to operate within a time cycle nearly that of the launch site. American space shuttle crews are launched on Eastern Standard or Eastern Daylight Times (zone -5 or zone -6 times) and operated in synchronization with mission control at Houston (Central Standard or Central Daylight Times, zone -6 and zone -7 respectively). Soviet and Russian crews have operated on Moscow time (zone +2). Time zones are roughly fifteen degrees of longitude wide and start at the meridian of zero longitude that passes through the Royal Observatory at Greenwich, England, now part of the London metroplex. Because of this, universal or Zulu Time (Z) was once called Greenwich Mean Time.

No reason exists to have space facilities operating on different terrestrial time zones. They can operate on any 24-hour time basis desired. It won't be very long before all space facilities adopt the time standard of worldwide aviation, Universal or Zulu Time (Z). All flight times and weather reporting times in the United States today use Universal Time.

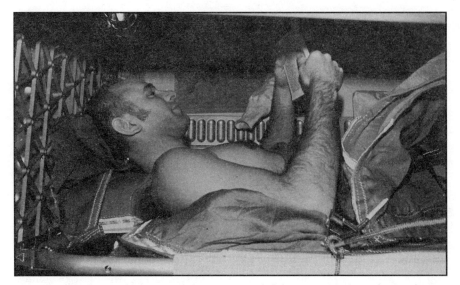

FIG. II-2: *Reading in bed in weightlessness. Was Skylab astronaut Alan Bean reading some science fiction space epic?* (NASA photo)

Human beings will take the 24-hour daily cycle into space with them because their internal clocks operate on a deep biological level. Experiments have been made with people living on different time cycles. These have shown that people can adapt to 20-hour and 28-hour daily cycles but operate best on the 24-hour cycle built into them by several billion years of evolution.

Work-Rest Cycles

A person's ability to work efficiently if proper work and rest periods are scheduled has led to some operational work-rest cycles used on Earth that can serve as patterns for those in space.

The US Navy uses a watch system known as "watch on watch." It's an emergency schedule of four hours on watch and four hours off watch. However, this 4-4 system can't be used for more than a week before performance deterioration becomes evident.

The NASA Skylab crews in 1973–1974 used a 16-8 system—16 hours of activity and 8 hours of sleep with all three crew members sleeping at the same time. However, astronaut Gerald P. Carr reported, "A guy needs some quiet time just to unwind if we're going to keep him healthy and alert up here. There are two tonics to our morale—having time to look out the window and the attitude you guys (Mission Control) take and your cheery words."

FIG. 11-3: *Although some people prefer to grow beards in space, Skylab astronaut Alan Bean shaved daily with a wind-up version of an electric razor.* (NASA photo)

In a highly industrialized culture—the only kind that permits space travel and living—any good foreman or manager can confirm these observations. In the early days of space exploration when only two or three people could be placed in space, it was necessary to get as much work out of the crew as possible because of the effort and expense of getting and keeping them there. This changed over the years. As better spaceships became available and routine, scheduled, economical space flights became possible. With crews of people living in space facilities for months at a time, the primitive working conditions of space exploration gave way to those of space habitation.

Normal daily space living follows an 8-8-8 sequence—8 hours of work, 8 hours of recreation, and 8 hours of sleep. At times because of unique situations this can go to 12-4-8, but there should never be less than 4 hours for "unwinding" and 8 hours of sleep unless an extreme emergency occurs.

Sleeping in Weightlessness

Sleeping in weightlessness is an experience in itself. Even the most primitive space sleeping arrangement can be more comfortable than using a water bed or even grandmother's legendary feather bed. A person can sleep anywhere because there is no need to get into bed or even to lie down.

The only physical requirement for sleep is restraint to keep a person from floating around and banging into the contents of the compartment because of constant air movement. Even a minute current of air movement is capable in a six-hour period of wafting an unrestrained sleeping person gently toward the screened air return duct with all the other free-floating objects in the compartment.

Sleep sacks are probably the solution to this. Built like an ordinary sleeping bag, a sleep sack can be attached to any surface of a compartment in any orientation. A person zips in, and the sleep sack prevents floating around in the air movement of the life support system.

Sleep sacks have been used in space since the days of Skylab in 1973.

Two cautions must be voiced about sleeping in a sack in weightlessness.

1. Although the human eyes and brain will identify "up" and "down" instantly within any compartment, the sleep sack should be positioned so that the person's head is "up" with respect to the way the compartment is perceived. Otherwise, some momentary disorientation may be experienced when the person wakes up from a sound sleep. The eyes will immediately confirm that the sleeping person is "right with the world" and thus less likely to suffer from a bout of space sickness. On the other hand, one Skylab astronaut preferred to sleep "head down" because he said that air blowing up his nose kept him awake.
2. No one should automatically assume that they'll be able to reach out in the dark and locate something like a light switch. Eye-hand coordination may adapt to weightlessness, but the kinesthetic sense may not. It was Skylab astronaut Owen Garriott who first reported that in the dark he was likely to miss by as much as 45 degrees in angle and 8 to 12 inches in distance. Therefore, a dim night or sleeping light may be helpful so that people don't suffer from a common nightmare: awakening to discover they're falling in the dark.

Some astronauts and cosmonauts reported sleeping in weightlessness was so restful that they required only 6 to 7 hours of sleep instead of the usual 8 that are necessary on Earth.

Eating in Space

Time not spent in working and sleeping in space must be devoted to what Gerald Carr called "unwinding."

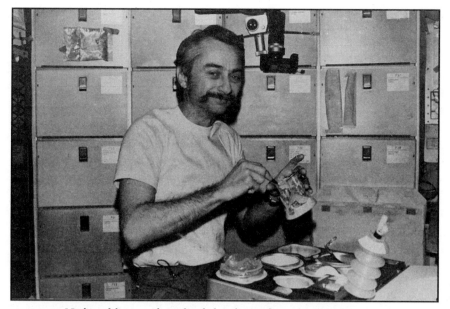

FIG. 11-4: *Nothing like a good meal to help relax and socialize.* (NASA photo)

This includes eating.

People need more time to eat in space. As the Skylab crews discovered, the three-meal day doesn't always work. During initial space adaptation, people haven't been able to go as long between meals. They become hungrier sooner and feel hunger pangs more intensely. Eating smaller meals four times a day helped, and snacking between meals helped even more (although such eating habits may be frowned upon in the "slim and slender" high-tech cultures on Earth).

The three-meal day is historically quite recent to human experience and isn't universally followed everywhere on Earth.

Our hunting ancestors ate when they felt hungry (if they had food available) or whenever they made a kill. The human body is fully equipped to function for several days without food. Fasting is usually followed by a period of gorging, scarfing down everything that can be eaten. An example of the human being's adaptation to this hunting existence is the presence of the gallbladder. This is a small pear-shaped sac adjacent to and partially attached to the underside of the liver. It connects through tubular ducts to both the liver and the portion of the intestinal tract called the duodenum just below the stomach. Its sole function is to receive and store the bile produced by the liver. Bile is essential for the digestion of fats and fatty substances in food.

The gallbladder stores enough bile to help digest a *big* meal such as that following the kill of a deer. However, the gallbladder is notorious for holding gallstones that are crystalline precipitations of cholesterol manufactured by the liver. Many people today have had their gallbladders removed because of gallstones, yet they continue to eat and digest food normally. Even without a gallbladder, gallstones continue to be formed and pass down the bile duct to the duodenum. The gallbladder is therefore a holdover from the days when our ancestors *didn't* eat regularly but had to gorge themselves when they made a fresh kill.

The habit of eating at a regular time thrice a day can be traced back to the court of the French king Louis XIV (1638–1715). Before that time in both castles and hovels, food was always available for anyone who was hungry. The court of Louis XIV not only developed tableware and table manners as we know them today but also the three-meal standard. So many people thronged the court of Louis XIV that the palace kitchens couldn't keep up with cooking meals for courtiers when they wanted to eat. Therefore, kitchen convenience is the only reason for the tri-meal day.

The three-meal custom may be followed in some space facilities that don't house many people and where the sociality of a meal is as important as the caloric content. Large facilities may operate on the 24-hour cafeteria system because most space foods will be ready-to-eat packages of individual portions available at any time from the food locker.

However, the social aspects of a meal mean that some people may decide among themselves to have regular meal hours if for no other purpose than to talk and engage in social recreation. This usually makes a meal taste better, as anyone who has had to eat alone on a regular basis will attest. Companionship and social interaction are as important in space as they are on Earth.

Exercise

Some recreational time must be devoted to exercise, and this is not because of any physical fitness fad. The human body slowly deteriorates or atrophies in weightlessness because its muscles and heart don't have to work so hard without gravity.

Plans call for the construction of large space facilities having centrifuged living quarters to provide the pseudogravity of centrifugal force that may be necessary for prolonged living in space. Otherwise, space facilities have mechanical exercising equipment and a definite requirement for a physical fitness schedule for all occupants. An hour of exercise daily will keep most people

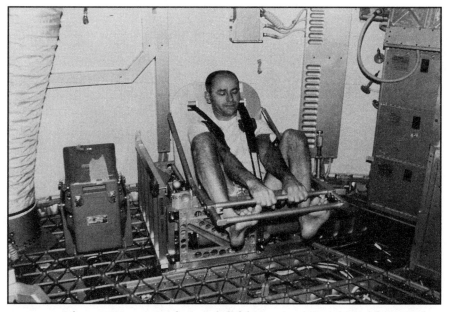

FIG. II-5: *A bungee-type exercise device is helpful in maintaining physical fitness and muscle tone in weightlessness.* (NASA photo)

in excellent physical condition and make it easier for them to re-adapt to the 1-g environment of Earth when they return.

Weight-lifting isn't a viable exercise in space because of the weightless condition of most facilities. Many exercise machines are based on pushing or pulling springs; these work in space as well as on Earth if they have body restraints to keep a person in place while tugging or pushing.

The simplest exercise machine is the ordinary "bungee" or elastic cord with a handle on each end. A bungee can be used on Earth or in space.

A bicycle ergonometer is useful in keeping leg muscles and the heart in good condition, although the rider has to be secured to the seat.

For those who like to run or jog, a treadmill is always a poor substitute, but it's the only way to get a good running or walking workout in weightlessness. Bungees or restraints must be used to keep the feet in contact with the treadmill.

In some of the larger facilities with diameters of ten feet or more and in open compartments that take in this whole diameter, it's possible to "run around" the entire inner circumference of the compartment. The Skylab astronauts were the first to do this. It's not running as much as it is exercise because a person must move fast enough to generate a little centrifugal force that will keep their feet against the inner circumference.

Astrobatics and Other Athletics

This leads immediately to zero-g acrobatics or, more properly, *astrobatics*. Whole new forms of both gymnastics and dance become possible in both weightlessness and the reduced pseudogravity of 0.1 g available in centrifuged facilities. The closest people can come to this on Earth is on a trampoline, a three-meter diving board, or underwater ballet; each has its drawbacks in the amount of time of falling or buoyancy. Astrobatics, on the other hand, is a good form of exercise not only for the body but the mind. It helps develop better kinesthetic abilities (knowing where the arms and legs are) as well as helping exercise the semi-circular canals of the inner ear. In their more mature forms, both astrobatics and dance can develop into new art forms as they become more standardized and sophisticated. Athletic competition using rigoristic formats are certain to be organized. In any field where two people can compare themselves, formalized competition has always arisen. Human beings are a competitive species.

Can we look forward to a weightless "space Olympic" gymnastic competition? Probably. Because of the excellent communications that exist between space facilities and Earth, it's likely to become a popular sport and a drawing card for space tourism because of its novelty.

Athletic games are possible in weightlessness. The existence of large open volumes in space facilities makes possible a whole new gamut of athletic competition.

On Earth, water polo could be played underwater with neutrally buoyant players and ball, the players wearing scuba gear. But because of the viscosity of water, the game is slow. "Space polo" is another matter. In a large weightless compartment with a goal located at either end and teams of players pitted against one another in the task of getting the ball into a guarded goal, this game is fast and three-dimensional. Basically, the rules state, "No kicking or gouging, biting, or clipping; no holding for more than two seconds; come out flailing and may the best team win." Football helmets may be required because, in such a weightless melee, it's possible to get kicked in the face or head. Other protection such as shoulder, elbow, and knee pads may also be required as the competition grows more intense. Throwing the ball itself results in the thrower exhibiting a reaction force unless there's a wall nearby. Personal motion is the result of pushing against a wall or an opposing player, which act causes the pushed player to go sailing off in the opposite direction. There are possibilities for injury involved with getting kicked, being involved in interpersonal impacts, and colliding with the walls.

In time, space polo televised from orbital facilities or even a Moon Olympics conducted under one-sixth lunar gravity may compete on terrestrial tele-

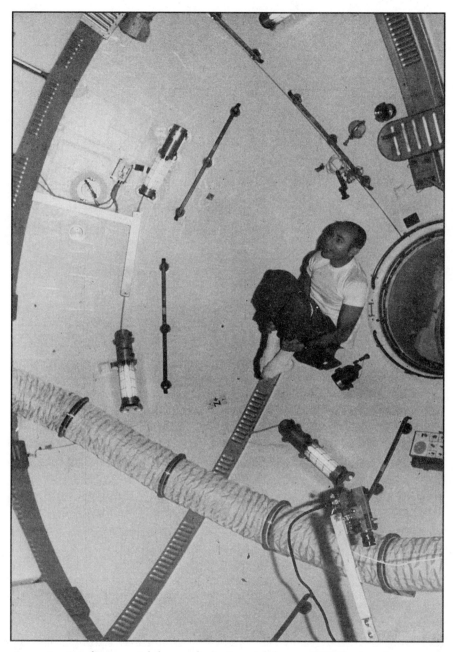

FIG. II-6: *Astrobatics, weightless acrobatics in weightlessness.* (NASA photo)

vision with professional football, basketball, ice hockey, wrestling, and boxing. Even handball, squash rackets, paddle tennis, and tennis itself become totally new games when played in the enclosed volume of a weightless space facility.

Special space sports facilities will surely be built for such games. They won't be any more expensive than the huge stadiums built all over Earth for sporting competitions. This has been going on for centuries, as the Roman Coliseum attests.

All of these space games are powerful, active, and even violent competitive sports because human beings with their hunter ancestry are competitive animals. As space sports grow in popularity, competitions will probably take place between teams from different facilities, even international ones, just as they take place today between sports teams from various cities. Such sporting activity defuses what many people believe is the greatest human hazard in space and on Earth: international rivalry leading to military confrontation.

Social Interaction in Space

All social interaction isn't competitive, however. There are competitive and noncompetitive social activities on Earth, and it's no different in space.

Human beings will take their social graces into space because the rules for acceptable conduct such as manners and protocol have developed over centuries and are integral parts of everyone's social mechanisms.

All social organizations have what the term implies: organization. The basic nature of a social institution is its internal structure. Space living offers some new opportunities in this regard. No matter where they are and what they're doing, people always need to know who's in charge and what responses are expected of everyone. This is simply a matter of manners. Since deep inside every human being is a territorial hunting animal, manners serve as a lubricant in interpersonal contacts. Every time people interact, they unconsciously tell themselves, "This is not worth fighting about, so today I'm not going to kill." Human covetousness and other aggressive traits derive from the evolution of *Homo sapiens* in a world of apparent scarcity where there never was quite enough to go around for everyone. In the twentieth century, people began to realize that this wasn't true because they lived in an abundant world if they applied their brainpower to make it so. But while some people were learning how to be rich and largely in control of the forces of nature, others were saddled with those ancient, ancestral traits developed to survive in a preindustrial world.

As with any area of human endeavor, a certain camaraderie develops among those who live and work in space. It's very much like that of sailors and aviators because all face common physical enemies. In space, the common physical enemy is always present behind the pressure bulkhead. Inside the pressure hull, people experience the same sorts of interpersonal contacts and the inevitable moments of confrontation.

This was confirmed even in the early days of space exploration. The Soviet cosmonauts reported that personal problems developed between their two-man *Salyut* crews during long stays in those space stations. Similar difficulties existed among some members of American astronaut crews but they were deemphasized by both the crew members and NASA.

People will party in space, especially the tourists (of course). Those social gatherings of a recreational nature allow people to do what they've done for millennia: get together to talk, eat, and play. Both Americans and Russians have celebrated holidays in space including crew-member birthdays. Such gatherings for celebration are an integral part of humanity's hunting heritage. They are absolutely essential for getting along together in space and on Earth.

Men and Women in Space

Early space exploration was male dominated. Although the first astronauts and cosmonauts had to be experienced experimental test pilots, several outstanding women such as Jacqueline Cochrane Odlum, Shiela Scott, and Jacqueline Cobb were equally qualified. For many years, Valentina Treshkova Nikolayeva was the only woman who'd been in orbit, riding the Soviet *Vostok 6* capsule on June 16, 1963. (She wasn't a test pilot, only a sport parachutist, but the Soviets were committed to being first in space at that time.) It was nineteen years later in 1982 that the Soviets put a second woman into orbit, Svetlana Savitskaya. The first American woman in space was Dr. Sally Ride who rode the space shuttle *Challenger* in 1983.

Women have been in space ever since and there will always be women in space from now on. This is as it should be. After all, space is more than a place where scientists pursue their favorite hobby: scientific investigation. It's a *human* activity. This was acknowledged as long ago as 1903 with Russian space pioneer Konstantin Eduardovich Tsiolkovski wrote, "Earth is the cradle of reason, but one does not remain in the cradle forever."

The integration of men and women in space naturally leads to the consideration by authors, space buffs, and even the public of the possibilities of "humanity's favorite recreation" in the weightless environment of a space facility. This is because humans are basically obsessed with sex, Dr. Sigmund Freud notwithstanding. This stems from the ancestral heritage of the human race where sexual activity was directly linked to reproduction; early hunters knew that animals did what they did to produce offspring, and this wasn't converted in a Great Mystery until the development of civilization began about 5000 B.C. The sexual drive is even stronger than the instinct for self preservation. Reproduction is the only way that immortal humankind can achieve any semblance of immortality in a mortal world.

The concept of romantic love grew slowly from this basic instinct. It accelerated as a result of the transition from a have-not hunting/agricultural civilization to a post-industrial world of abundance. This in turn has led to a re-evolution of sexual attitudes. (In the ancient cities of Sumer five thousand years ago, men and women were treated as equals.) As the world of abundance evolved over the past several centuries, sexual activities became more and more recreational and, in some cultures, even an art form.

Of course human beings will take sexual activities into space with them! They're doomed if they don't and, according to some prudes, damned if they do. However, if in the long run human beings wish to colonize space, men and women will have to reproduce themselves.

The first actual space colony won't be built; it will be *created* in any space facility when the first child is born there.

As of this writing, human sexual activity in space hasn't *openly* happened yet. There have been rumors and stories, but no confirmation.

This doesn't mean that people haven't been thinking about it. Back in the 1980s, some clandestine experiments were conducted very late at night in the neutral buoyancy weightless simulation tank at NASA's George C. Marshall Space Flight Center in Huntsville, Alabama. The experimental results showed that yes, it is indeed possible for humans to copulate in weightlessness. However, they have trouble staying together. The covert researchers discovered that it helped to have a third person to push at the right time in the right place. The anonymous researchers—who would have been fired if their activity had been revealed—discovered that this is the way the dolphins do it. A third dolphin is always present during the mating process. This led to the creation of the space-going equivalent of aviation's Mile High Club known as the Three Dolphin Club.

Many people would like to go into orbit to try this new recreation. The Japanese are well-aware of this. In the 1990s, they conducted several market studies of space tourism among people in the United States and Canada. As a result, they have identified their primary market target for space tourism: honeymoon couples.

Therefore, gravity isn't necessary for procreation. The neutral buoyancy tank experiments proved that humans hadn't forgotten what their very remote ancestors learned in the oceans of Earth. There, species learned how to copulate while swimming in a neutral buoyancy environment. Those species survived and are the distant ancestors of humanity.

Therefore, in space it's just a matter of doing what comes naturally because, in essence, people are returning to the neutral buoyancy condition of Earth's oceans.

People must rely on personal experience to a large extent in considering this because everyone has had a definitely personal experience in this area that's unlike that of anyone else. Sexual activity is unique in human experience because, in spite of books and even personal coaching, no one really knows anything about it until they've done it.

Therefore, no reason exists to prevent sexual activities of many sorts from being adequately and satisfactorily carried out in weightlessness. Thanks to humanity's ancestors who lived in trees, everyone is equipped with arms, legs, and hands that can grasp. Therefore, any action and reaction forces can, with some experience, be controlled, probably without the third person being present.

A sleep sack is probably the best place for an initial encounter. A modest level of physical restraint exists to prevent uncontrolled forays around the compartment. Padded compartments in some space facilities may cater to tourists who want to be free to experiment.

Judging from the plethora of ancient temple carvings as well as modern printed manuals, not a single one of the legendary "Thousand-and-one Ways of the Ancients" won't work in weightlessness.

It's not necessary to be explicit. Nor is there any reason to be apprehensive. People will do well. Fish do, and they don't have arms, hands, and legs.

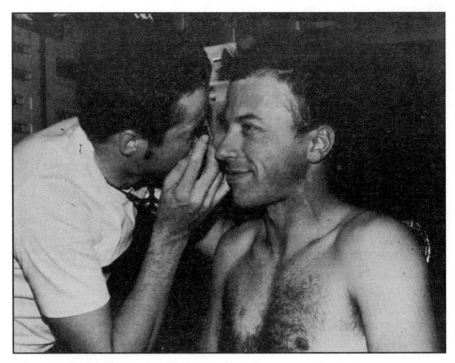

FIG. 12-1. *Dr. Joe Kerwin was the first medical doctor to practice in space.* (NASA photo)

CHAPTER TWELVE

HEALTH AND MEDICINE IN SPACE

Preliminary Medical Requirements

In the foreseeable future, people won't be allowed to go into space to live and work unless they're in good health and suffer no serious congenital ailments. However, as the number of people living in space increases and as more information and experience is acquired, many of the stringent medical requirements initially imposed will be relaxed. This has been the general trend in space living. Of course, space tourists need meet far less strict health requirements if any at all.

The first seven Mercury astronauts had to be in top physical condition. The medical examinations given to these super-healthy test pilots at the Lovelace Clinic in Albuquerque, New Mexico, were perhaps the most extensive ever devised because of the unknown physical effects of the space environment. As these concerns disappeared because of growing space experience, the medical qualifications and standards were relaxed. By the time the NASA space shuttle began flying in 1981, the required medical exam was no more difficult than that for an FAA Third Class Airman's medical certificate. This is passed every two years by three quarters of a million private pilots in the United States.

In the past, astronauts have lost their chances to participate in space missions because of temporary minor illnesses or possible exposure to contagious communicable diseases. However, even with these precautions, the three-man crew of the week-long Apollo 7 mission all came down with head colds after getting into orbit. This led their waggish commander, Wally Schirra, to call Apollo 7 the "Seven Day Cold Capsule."

Basic Medical Requirements

The medical requirements for space travel will probably follow those of the current Federal Aviation Administration standards of physical eligibility for a Third Class Airman's Medical Certificate. All student and private pilots must meet these requirements. Those for commercial and airline transport pilots are far more stringent. These requirements are:

Eyes. Distant visual acuity of 20/50 or better in each eye separately without correction or, if corrective lenses are worn, 20/30 or better. No serious pathology of the eye.

Ears. Ability to hear the whispered voice at 3 feet and no acute or chronic diseases of the inner ear. No disturbance in equilibrium.

Mental and neurologic. No established medical history or clinical diagnosis of any of the following:

1. a personality disorder that is severe enough to manifest itself by overt acts;
2. a psychosis;
3. alcoholism, meaning a condition in which alcohol has become a prerequisite to a person's normal functioning;
4. drug dependence, as evidenced by habitual use or a clear sense of need for the drug;
5. epilepsy;
6. a disturbance of consciousness without satisfactory medical explanation of the cause;
7. a convulsive disorder, disturbance of consciousness, or neurologic condition that might endanger a person or others.

Cardiovascular. No established medical history or clinical diagnosis of myocardial infarction, angina pectoris, or other evidence of coronary heart disease that may reasonably be expected to lead to myocardial infarction.

General medical condition. No established medical history or clinical diagnosis of diabetes mellitus that requires insulin or any other hypoglycemic drug for control. No other organic, functional, or structural disease, defect, or limitation that might affect a person's ability to perform safely the activities of living in space.

Naturally, there are individual cases in which various sorts of waivers will be granted. Any medical standards this generalized can't be as closely applied as those of the First Class Airman's Medical Certificate required for airline transport pilots. Individuals can and do deviate from these general norms and are otherwise considered healthy and capable.

For example, many private pilots have obtained waivers for the cardiovascular requirements because, although they've suffered heart attacks, they've undergone bypass surgery or have undertaken rigorous rehabilitation programs to recondition their hearts. Handicapped pilots such as those belonging to the Wheel Chair Pilots Association have obtained the necessary waivers to fly even though they may be, for example, partially paralyzed.

As the years go by, more and more waivers from the medical standards and requirements will be given. Space conditions themselves may turn out to be benign for certain handicapped people.

Communicable Diseases in Space

If a person is suffering from an acute or chronic communicable disease, however, it's going to be a different story. Space facilities are the healthiest of all human living situations and must be maintained that way for several reasons.

In the first place, a person is in space to do a job that a machine can't because a human can do it better and cheaper. If that person becomes ill and has to take sick leave in the facility, the cost of maintaining that individual's life support continues during the period of illness or incapacitation. Contingencies such as temporary illness are taken into account during the design and crewing of the facility so that a few people on sick call won't disrupt daily operations.

A similar operational philosophy exists aboard ships on the Earth's oceans and in the frontier living conditions of the North Slope of Alaska or the scientific stations in Antarctica.

It becomes a serious matter, however, when a communicable disease triggers an epidemic in a space facility.

If an epidemic breaks out in Prudhoe Bay or Byrd Station, help is only a few hours away by air if the weather cooperates. The sick can be flown out or immunizing agents and antibiotics flown in. The analogy holds true for most of the space facilities in low Earth orbit but may not hold for those in geosynchronous orbit and certainly not for those in lunar orbit or on the Moon. The greater the distance from the Earth or from a major space medical facility, the longer it takes to get help or to move the sick and injured to better facilities, if such facilities exist and if the sick and injured can be moved.

This is why extreme measures must be taken in the early years of space living to prevent communicable diseases from getting into space facilities.

Quarantine

To gain entry into some space facilities, especially those involved in the manufacturing of biological materials where exceptionally stringent standards have been established, a person may be quarantined for several weeks until it can be ascertained that no undesirable microorganisms are present in that person's body, the intestinal flora have been flushed and replaced with those more benign to the biological operations taking place in the facility, or the person can be certified to be physiologically and pathologically "clean" with regard to the activities taking place there.

Other facilities, especially those having total or partial military functions, may require quarantine to prevent the covert introduction of chemical or biological agents. Space facilities are exceedingly vulnerable to a wide variety of these.

Space-Made Pharmaceuticals

However, because of biological materials made in the weightlessness of space, it's very likely that immunizations and specifics for nearly every known disease, including the common cold, will be available. These space pharmaceuticals will ensure that space facilities remain the healthiest of all human environments and will do much to stem the tide of sickness and disease on Earth where, because of the larger biosphere, it's more difficult to control the total environment.

The problem of communicable diseases in space will be solved by biological materials made in space.

Personal Hygiene

However, personal hygiene will continue to be important in space facilities regardless of whether or not biological specifics become available for most communicable diseases. The military services long ago discovered that personal hygiene is extremely important when groups of people live and work in proximity to one another.

Personal hygiene means that an individual must keep clean and keep the surroundings clean as well. This doesn't necessarily mean sterile clean, but hygienically clean.

Any medical pathologist will confirm that the human body is a breeding and feeding ground for a wide variety of microorganisms, some of which are pathological and others of which are benign. People live in a symbiotic relationship with thousands of other terrestrial organisms. Sometimes organisms that are symbiotic, harmless, or helpful to one individual can be hazardous to others. Earthbound hospitals are extremely aware of this fact, although they face a herculean task of handling an enormous number of microorganisms and aren't always successful in doing so.

Microorganisms and viruses enter an earthbound hospital in the air, in supplies, and on people entering the building from the outside world. In space, the only source of microorganisms will be incoming supplies and people. Methods exist to reduce the influx of organisms in and on supplies, but there is no way to prevent them from coming in on people because everyone is alive and crawling with such organisms.

In addition to the combination of natural human microorganisms are the odors that people exude as part of perspiration. This means that a space facility will begin to smell like an athletic locker room or worse if people in it don't maintain good individual and collective hygiene.

Because of the requirements for good personal hygiene, space facilities make most earthly habitats seem incredibly dirty in comparison.

Personal Hygiene Kits and Activities

The design and function of space facilities or the nature of some spaceships and flights may require that people be provided with a personal hygiene kit such as the ones pioneered in the NASA space shuttle program. A PH kit for men contains, for example, such material as a razor, comb, brush, toothbrush, nail clippers, nail file, styptic pencil, and a week's supply of shaving cream, skin cream, stick deodorant, dental floss, antichap lip balm, and soap.

The operational requirements of long duration space facilities with advanced life support systems may allow people to make up their own personal hygiene kits from a list of commercial products that have been selected for their compatibility with the life-support system. They also may be allowed a limited selection of other personal items.

A person in space must stay personally clean. Personal hygiene measures in space are no more stringent than those necessary when living in any close quarters on Earth with the exception that an individual can't get away from the space facility on a moment's notice. In this regard, submarines are a close analogy.

A person should wash the face and hands regularly and take a shower as often as permitted and possible. If taking a shower isn't possible, a sponge bath will suffice. People should not depend upon antiperspirants and perfumes to knock down personal odors; these don't work very long and usually restrict perspiration that may be needed to stay cool.

Clean clothes must be worn and when clothing gets dirty it should be cleaned. Doing laundry in space may be out of the question in small facilities that may not have the equipment for doing this. Clothes may have to be sent to other facilities or back to Earth to be cleaned. This is what the forty-niners in San Francisco did. In 1849, dirty laundry was actually shipped by sailing vessel all the way from San Francisco to Hawaii to be washed and returned because San Francisco didn't have a suitable water supply or the people available to do laundry.

Waste Management Hygiene

People who have recently entered the zero-g environment of a space facility may suffer from occasional bouts of motion sickness that may cause vomiting. Therefore, sick sacks and packaged moist towelettes are items that wise space workers carry with them. People new to space quickly discover that they have to clean up the mess themselves because everyone else has other work that must be done. Liquids go all over in weightlessness and are difficult to control once they get loose in a facility. So the clean-up job has to be done at once, unpalatable as it may be.

The most useful item for controlling both liquid and solid waste matter is the disposable moist towelette. These are available in sealed packages that people can carry with them. Used towelettes should be disposed of properly and promptly in the proper waste management receptacle.

It's equally important to maintain strict personal hygiene when using the various types of zero-g toilet facilities. Most of these operate on the same principles as those pioneered in Skylab and the space shuttle. The force of gravity is replaced by a flow of air to draw urine and feces into collection enclosures. The more advanced life support systems reprocess urine by extracting the water from it and storing the residue as solid waste. Fecal collectors use a similar principle to separate water from the solid matter. Antibacterial agents are added to inhibit the growth of bacteria and eliminate the possible release of odors. Solid wastes from urine and feces may be vacuum-dried in containers that are replaced from time to time with new ones while the used ones are returned to Earth or a space disposal facility. In the more advanced closed-cycle life-support systems, solid waste is recycled by the system into nutrients for the space farms that not only help recycle carbon dioxide into breathing oxygen but also provide food for the facility.

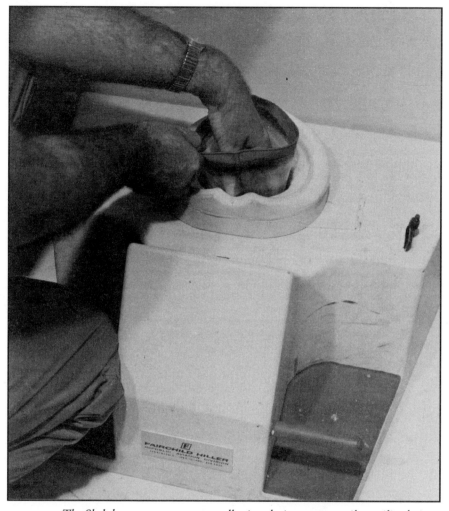

FIG. 12-2. *The Skylab waste management collection device, a space toilet, utilized air movement to pull wastes into a disposable plastic sack that was later removed and sealed. Current space toilets have eliminated the need for the removable sack.* (NASA photo)

Living in some remote space facilities or spaceships without dedicated toilet facilities may require that people use an up-dated version of the old Gemini Bag.

Some people have trouble adapting to the techniques of personal waste management in space because of cultural and toilet training on Earth. In space, they must overcome the distaste or other psychological blocks they may have; if they don't, they find living in space to be very difficult.

Waste Management in a Pressure Suit

Dealing with waste management in a pressure suit (space suit) poses different problems.

Pressure suits have built-in urine collectors. When a person "suits-up," it's important to ensure that either the male or female urine collector is securely attached so that there's no possibility of leakage. When desuiting, one must be careful to seal the sphincters and close the valves of the collection system to present the escape of wastes. In some facilities, space suit technicians take care of removing the collection systems and disposing of any material in them. However, in most facilities, this task is one that everyone does for themselves.

If a job requires many hours in a pressure suit every day, people may develop personal diets and hygienic regimes that permit them to spend long periods suited-up without having to relieve themselves. However, space suit waste management systems should *always* be properly attached and available in case of an emergency. When a person is wearing a space suit, it's practically impossible to attach the waste collection system if it wasn't hooked up during suit-up.

Space Medicine

Although space medicine was recognized as a definite branch of the medical profession long before people first flew into Earth orbit, it was essentially an offshoot of aviation medicine. As such, it was primarily concerned with human physiological responses to the aerospace environment. True space medicine—the prevention and treatment of injuries and illness in space—didn't start until people began to live in space for more than a few days at a time. Now space medicine is an active area of the medical profession. Physicians have practiced medicine in space and will continue to do so in the future as more and more people live in space to carry out an increasing variety of scientific, technical, industrial, construction, and support tasks.

There are eight basic areas of space medicine:

1. Trauma medicine concerned with job-related or accidental physical injuries such as cuts, burns, bruises, abrasions, broken limbs, and even partially or totally severed limbs.
2. Pathology, especially that concerned with personal hygiene, "public health" in a spaceship or facility, and preventive and anti-epidemic measures dealing with infectious agents such as bacteria, fungi, and viruses.

3. Treatment or amelioration of suffering related to congenital afflictions such as appendicitis, tonsillitis, cholecystitis, earaches, toothaches, and allergic reactions to some types of materials used in spaceships and facilities.

4. Treatment of stress-related illnesses manifesting themselves in hypertension, cardiovascular problems, and even psychosomatic conditions.

5. Handling of psychological problems caused by the unnatural environment of a space facility and phobias that may not manifest themselves until a person has been in the space environment for several hours or days. These might include claustrophobia or acrophobia, afflictions that surfaced even among the early astronauts who were highly trained and motivated.

6. Biochemical problems brought on by dietary deficiencies, glandular imbalances, blood chemistry problems, lack of trace elements in food, etc.

7. Medical problems caused by social interaction. Conflicts will occur regardless of how well people appear to get along with one another and regardless of rules, regulations, and security measures. Someone is certain to smuggle alcohol or drugs aboard to liven-up what they consider deadly boredom.

8. The basic medical problems unique to the space environment itself. These include hypercalcemia due to calcium resorption, the Kittinger Syndrome or "vac-bite" caused by accidental exposure of part of the body to vacuum; pneumo-poisoning caused by an excess of oxygen, nitrogen, carbon dioxide, and carbon monoxide; and even "traumatic abaria" or the partial or momentary loss of ambient pressure. Included among these space-related medical problems are the physiological and psychological effects of ionizing radiation, especially in those facilities located beyond the Van Allen belts where ionizing radiation from the Sun becomes a serious health hazard.

Contingency Medicine or "First-Aid"

The portion of space medicine people are most likely to encounter on a personal basis has been termed (in the usual aerospace parlance of ellipsis and equivocation) "contingency medical activity." It's easier to call it by its older, well-known name: "first-aid."

Because physicians and medical facilities may not be available in many spaceships and in some of the smaller facilities, and because hours or even days may pass before professional medical help can be obtained, people will be thoroughly trained in first-aid techniques for minor illnesses and accidents, especially to stabilize the physical condition of someone who's been severely injured.

In common with airliners, all spaceships will carry first-aid kits containing materials for emergency medical treatment and the relief of suffering for minor illnesses and injuries. Flight crews of airliners are already trained in the proper first-aid techniques.

First-aid kits are usually located in critical places in space facilities such as at or near air locks or near equipment whose malfunction could cause personal injury. All space facilities have either an area specifically set aside as an infirmary or hospital (depending upon the size of the facility) or an area that can be quickly converted into a medical treatment or infirmary compartment. Part of the initial orientation of everyone in any space facility is a briefing on the location of first-aid kits, where the sick bay is located, and the communications procedures to be used in calling for help from medical personnel in the facility or elsewhere.

All first-aid kits contain more or less the same materials, although more complex kits may be used in long-flight spaceships or remote space facilities not having dedicated medical capabilities. There may be as many as three plainly marked packages in a first-aid kit:

1. A medications packet containing oral, injectable, and non-injectable medications. Some of these may not be administered without the specific approval of a paramedic and, in some cases, a physician. However, in situations where a paramedic, nurse practitioner, physician's assistant, or physician isn't immediately available but contact can be made by radio, oral instructions for use of such medically restricted items may be given and considered as approval or permission.

2. A bandage and equipment packet containing Band-Aids, Steri-Strips, tape, gauze, wipes, swabs, injectors, syringes, intravenous injection equipment, tweezers, forceps, scalpels, scissors, hemostats, sutures, needles, sponges, and splints. The contents of any given kit will depend upon where it's located and the availability of medical personnel and facilities. Some kits may contain equipment that cannot be used except by professionals or that may be used by non-professionals provided communications are established with their approval or oral supervision.

3. A diagnostic/therapeutic packet containing a stethoscope, blood-pressure cuff and sphygmomanometer, airway, thermometers, penlight, catheters, otoscope specula, ophthalmoscope, binocular loupe, sterile drape, sterile gloves, and defibrillator. Here again, the type and nature of the equipment in this package depends on where the first-aid kit is located.

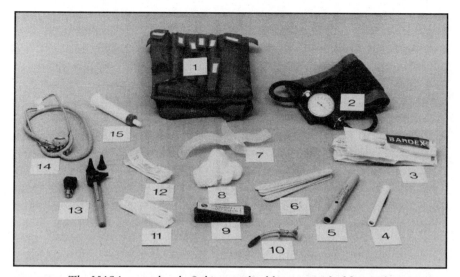

FIG. 12-3. *The NASA space shuttle Orbiter medical kit is typical of first-aid equipment found in spaceships and space facilities. Key: (1) bag, (2) blood-pressure cuff, (3) sterile drape, (4) flashlight, (5) disposable oral thermometers, (6) tongue depressors, (7) insertable airway, (8) cotton balls, (9) tourniquet, (10) Foley catheter, (11) sterile gloves, (12) fluorescein strips, (13) otoscope and ophthalmoscope heads, (14) stethoscope, and (15) lubricant jelly.* (NASA photo)

The use of some of the equipment must be carried out by or under the supervision and with the approval of qualified medical personnel.

All first-aid kits must have a complete list of their contents plus a quick-reference handbook giving explicit instructions for the administration and use of each item in the kit. Those medications and items whose use requires professional presence or remote oral approval are sealed in packs that are clearly marked regarding these instructions.

The complex first-aid kits containing certain drugs and equipment are the responsibility of a named person in the ship or facility who regularly checks the kit inventory. If seals are broken or items missing, the person responsible for the kits must account for the drugs and equipment taken from the kit.

First-Aid Techniques

Not only is no reasonable way available to prevent all pathogenic agents from entering a space facility, but work anywhere cannot be performed without the possibility, however remote, of an accident leading to an injury on the job. No matter what engineers do or how clever they are, they cannot design anything that will protect everyone. It's impossible to protect the world from idiots or to protect idiots from themselves. The only practical approach that can be taken within the bounds of reasonable costs, efficiencies, and reliability is to design equipment, tools, and other devices that are difficult to operate incorrectly. Managers and personnel evaluators must then hire people who possess the intelligence, discipline, and expertise *not* to get hurt in the first place. Space is no place for an accident-prone person.

However, in spite of the finest engineering and all the protection technologies required by regulations as well as management, people get hurt and will continue to get hurt on the job. Space is no exception because it's just another work environment. Every engineering job and industrial operation on Earth proceeds with an estimate of anticipated injuries and even deaths; it's no different in space.

People must be prepared for accidents. Some of these will be *unusual* accidents because, in spite of all the research and development that has gone into space living, space is still a hostile environment and one of the most deadly yet broached by human beings.

First-aid techniques in weightlessness are unique and cannot be duplicated or taught on Earth. Therefore, most first-aid training will be conducted during orientation in the space facility itself.

For example, intravenous injections cannot be administered by the drip method in weightlessness because no gravity exists to provide the force to transport the fluids. Positive displacement IV techniques such as those often used in earthbound hospitals must be used instead. Some of these IV injectors are manually operated and require very careful operation to ensure that the intravenous injection is properly metered into the patient. Other units are battery-operated and can be adjusted to provide a calibrated injection rate.

It's extremely important that people learn how to insert an airway quickly when administering first-aid to an unconscious person in weightlessness. To an even greater extent than it is on Earth in a 1-g environment, it's possible or even highly probable that an unconscious person has either swallowed his tongue or inhaled regurgitated stomach fluids. The rapid and timely insertion of an airway is therefore extremely critical *whether or not an airway appears to be needed at the time.*

Space Medical Facilities

Large space facilities and those located in geosynchronous orbit or beyond will usually possess well-staffed and completely equipped hospitals. Some of these may be small and thus deserve only the title of "infirmary." However, even the smallest and most austere space hospitals possess the basic equipment for diagnostic purposes as well as facilities for carrying out the most complex and advanced surgical techniques.

Space hospitals also have excellent communications with other facilities and with Earth where specialists and large medical data bases exist. No matter what the symptoms of affliction, the space medical staff can obtain consultations with other physicians through interactive television conferencing. They also have at their fingertips the amassed and indexed medical knowledge of the world in earthly computer data bases. Given the basic diagnostic data from the computers in the space hospital, the computer data bases, with the help of doctors in space and on Earth, can provide analysis as well as diagnosis, prognosis, and suggested avenues of treatment.

Surgical facilities located in centrifuged modules of large space stations can utilize techniques and procedures that differ little from those used on Earth. However, zero-g surgery is a highly specialized field that requires unique methods and equipment. Some surgery is easier to perform in weightlessness. However, surgeons and surgical nurses must have special training to use these zero-g procedures.

Postoperative, recuperative, and other hospital facilities requiring a high level of asepsis are possible in space facilities. This leads to a reduced use of broad-spectrum antibiotics that have serious side effects.

Summary

When all aspects are considered and even the high level of potential hazard in space are taken into account, space is one of the safest and healthiest of all environments when it comes to living and working there. Everyone must practice a high degree of personal hygiene because of the requirements to maintain a clean, operative life-support system. Every flight crew and nearly all persons living and working in space are trained in first-aid, and emergency first-aid kits are well distributed in spaceships and space facilities. The excellent communications and computer data bases available mean that medical assistance and complete medical libraries are almost instantly at hand. Because of the remote locations of some space facilities in terms of the time necessary to return a sick or injured person to space hospitals or Earth, medical facilities in space are among the very best available anywhere. Space is indeed a dangerous place in which to live, but it is at the same time the healthiest.

FIG. 13-1: *Spaceplexes may take on many shapes, and these constantly change as more modules are added.* (Art by Sternbach)

CHAPTER THIRTEEN

SPACE COMPLEXES

THROUGHOUT THIS BOOK, places where people live and work in space haven't been called "habitats," although that word is a perfectly good description of them in spite of projecting an image that they're intended only for people to live in. Nor have they been called "space stations" because that term conjures up images of primitive experimental outposts such as Skylab, *Mir*, and even the doughnut-shaped *Collier's* space station proposal of 1952. The word "facilities" has been used instead. However, even this word can be misconstrued. After some study, the word "complex" has been chosen to tag henceforth the space stations, space facilities, space habitats, and other space structures in which people live and work. Many types of complexes already exist on Earth. Even the term "metroplex" has been invented to describe a very large collection of cities and towns.

In fact, think of a space complex or "spaceplex" as the equivalent of a village, town, or city. These collections of structures and people on Earth exist for multiple purposes. They are different because of their locations, purposes, and functions. They have individuality. It's no different in space where spaceplexes may include space factories, space hotels, space warehouses, space filling stations, space power plants, and spaceship docking, fueling, repair, construction, and maintenance bases.

Military space complexes should be considered as extensions of their counterparts on Earth. Many military forts evolved into towns and cities because (a) they were in defensible locations, and (b) they commanded nodes along trade routes. Many American cities originated from strategically located military outposts sited to protect local inhabitants and those traveling past. They

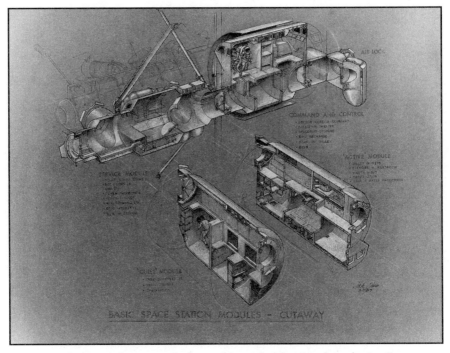

FIG. 13-2: *Cutaway of a proposed advanced spaceplex showing the radiation "storm cellar" as part of the command and control module.* (Boeing Company)

proclaim their heritage with city names begining with the word "fort"—Fort Wayne, Fort Worth, Fort Pierce, Fort Myers, Fort Dodge, Fort Collins, Fort Payne, Fort Smith, Fort Lauderdale, and Fort Lee, for example.

However, although some sort of space police force or military organization will be necessary to enforce space law and to provide the necessary physical coercion to back up ruling of space courts in resolving controversies, most people will live and work in spaceplexes devoted to scientific research, industrial production, or other commercial purposes. Each will develop its own social institutions and organizations just as earthbound cities evolve their own civic personalities.

The Two Basic Spaceplex Types

Some spaceplexes may be new, others old. Some may be primitive and others may offer a life style that many space pioneers would consider luxurious. Some

FIG. 13-3: *Military spaceplexes are necessary to provide security protection for valuable commercial facilities.* (Art by Sternbach)

may be small and intimate while large ones have a definite cosmopolitan feeling about them. Many may be international in nature with modules built and operated by various national crews while others are "pan-national" types catering to the needs of the space-going versions of multinational corporations, consortiums, and joint ventures. Some are "space industrial parks" with living quarters attached.

Except for some scientific and military outposts or national space facility modules, few spaceplexes are owned and operated by terrestrial governments. Like some airports and spaceports on Earth, the docking sectors of some spaceplexes may be owned and operated by a "municipal" spaceplex government. History has clearly shown that wholly-government-owned facilities on Earth and in space get mired in bureaucracy, are dependent upon the whirling knives of politics, lead to government monopolies, restrict initiative, and are operated under stifling controls. Early spaceplexes may be government-owned but only as long as it takes to convince business people, financiers, and bankers that risks are acceptable and money can be made building and operating spaceplexes. The basic core of a spaceplex and its government may resemble those of earthly cities.

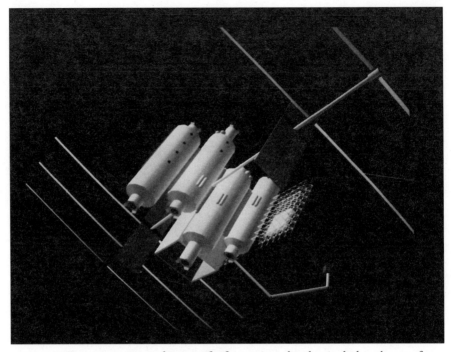

FIG. 13-4: *Computer-generated image of a future spaceplex that includes a hangar for servicing spaceships.* (Boeing Company)

Early space stations didn't have to offer a comfortable, almost terrestrial life style in order to entice people to come there to live and work. Most people who crewed these early spaceplexes were highly motivated to get into space to conduct scientific research, industrial experimentation and development, or commercial ventures. Other than requring a space outpost to provide them with basic life support, these pioneers didn't demand comfort. They were far more interested in safety, reliability, and functional characteristics.

However, as the new industrial revolution in space progresses, similar advances will be made in the evolution of unique life styles leading to more comfortable places to live. Here again, this is the way frontiers develop. The first towns are quite crude and small, and life on the frontier is difficult. This changes rapidly as the frontier develops.

Two basic spaceplex types may be found. One of these is the purely weightless spaceplex where all living and working takes place in zero-g. The other is the rotating spaceplex where centrifugal force produces pseudogravity for various human and industrial purposes.

Some may be combinations of these two basic types.

FIG. 13-5: *The "Pilgrim Observer" was a 1969 concept of a rotating space station that could have been used for planetary exploration as well. Its three rotating arms contained a nuclear power station, a crew living module, and a garden or "space farm" to recycle the atmosphere. It was available as a plastic kit and was one of the most predictive concepts of its time.* (G. Harry Stine)

Rotating Spaceplexes

Space facilities with rotating modules are far more difficult to design, build, and maintain. Therefore, they are built only because of the nature of the life-support system used, the stay times of people who live and work there, or the particular needs of the primary industrial user whose process requires the pseudogravity of centrifugal force.

Some advanced closed-cycle life-support systems use plants to recycle carbon dioxide into breathable oxygen by photosynthesis. These systems also provide some or all of the food of the occupants. Although plants have been genetically developed for growth in weightlessness, many plants that people want to grow in spaceplexes exhibit a strong geotropism, the inherent tendency for the roots to go downward and the stalks upward. For some plants gravity is needed to prevent reduced growth, delayed growth, and inability to fructify to produce seeds. As the years go by, more and more plants will be engineered to grow in a weightless environment. However, a plant-based life-support system isn't the only reason that a rotating facility may be necessary.

Some industrial processes may require a pseudogravity force because, at some point in the processes, density differences may have to be used for liquids or the liquids themselves may be extremely difficult to contain.

Distances, propellant requirements, locations, or other physical factors may make it too expensive to change crews regularly to prevent long-term physiological changes caused by zero-g. Long stays in weightlessness may make it necessary to centrifuge at least the living quarters to prevent physiological changes that may result from stays of several years in weightlessness. No "show stoppers" have been discovered thus far for people living in weightlessness for about a year, but little is known about the results of living for several years in zero-g.

The calcium resorption problem or the atrophy of the cardiovascular system may make it necessary to provide pseudogravity for long stays in space.

Living in a rotating spaceplex is like being aboard a ship. There are floors, ceilings, walls, doors, tables, chairs, benches, beds, food that doesn't float around, liquids that don't get out of control, toilets that work like those on Earth, and all the ordinary comforts of home.

However, there are some important differences as well.

The Coriolis Force Problem

Rotating spaceplexes have to be *large* in comparison to weightless ones because of something seldom encountered on Earth. Few people therefore understand it.

This factor is *Coriolis force*, named after the French mathematician Gaspard Gustave de Coriolis (1792–1843) who first described it.

It's a consequence of Newton's First Law of Motion: "An object at rest will remain at rest and an object in motion will remain in motion in a straight line, until acted upon by an external force."

Coriolis force is illustrated in Figure 13-6, which represents a rotating table seen from above. A line is drawn from the center to the outer edge of the table while it's rotating. To an outside observer, the marker appears to travel in a straight line from the center to the edge.

If a person had been riding on the rotating table when the mark had been made by the outside observer, it would appear that some force deflected the marker from a straight line and made it curve. The "force" isn't really a force at all, although it's called Coriolis force.

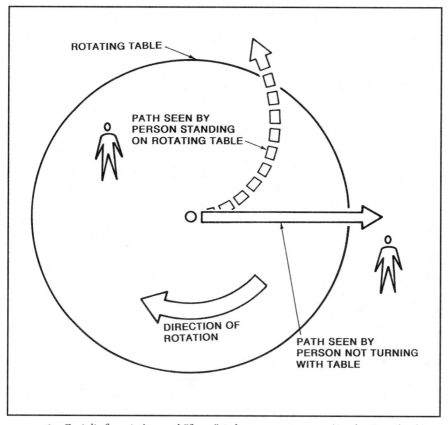

FIG. 13-6: *Coriolis force isn't a real "force," only an apparent one.* (Art by Sternbach)

On Earth, Coriolis force makes most winds blow from the west in the northern hemisphere and from the east in the southern hemisphere instead of from the north and south as they would on a non-rotating Earth.

How does this affect people living in a rotating spaceplex?

Figure 13-7 represents a rotating spaceplex shaped like a doughnut and called a "torus." As it rotates, it creates centrifugal force that holds the human on its inside surface as shown, creating pseudogravity.

Suppose the torus is 1,000 feet in diameter and rotating at 1.71 revolutions per minute to create 1-g on its inside surface. If the person jumps off a stool 21.5 inches high, Coriolis force will cause the landing to be displaced 2 inches to the side of the aim point.

Coriolis force has an effect on people, too. In a spaceplex rotating at several rpm, simple movements become complex and strong visual illusions are created. A rapid turning of the head can make stationary objects appear to gyrate and continue to move once the head is no longer turned. Coriolis force also creates cross-coupled accelerations of the fluid in the human balance organ, the semicircular canals of the inner ear. This in turn can lead to severe motion sickness.

Experiments on Earth have shown that people can probably adapt to rotations as fast as 3 rpm, but it takes training.

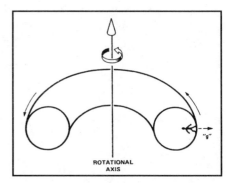

FIG. 13-7: *In a rotating spaceplex module, a person's head is closer to the axis of rotation than the feet and will therefore be subjected to less pseudogravity created by centrifugal force.* (NASA)

Human Tolerances to Rotation and Gravity Gradient

Rotating spaceplexes must be large for other reasons, too.

The first has to do with human tolerance to rotation. Everyone has experienced the dizziness that results from rapid rotation, so this isn't a new factor. Again, this was initially encountered in aerobatic airplanes and later in uncontrolled gyrations experienced by parachutists.

The US Air Force human factors researchers encountered human tolerance limits to rotation when testing the first jet airplane ejection seats. Some of these would tumble rapidly through the air after coming out of the airplane cockpit. The rapid rotation caused the seated human to lose consciousness. Therefore, they set about to learn about human tolerance to rotation. Some of the early standards were somewhat arbitrary. The US Air Force was persuaded to increase their rotational tolerance standards when a research group of which the author was a member measured the rotation of an ice skater doing a spin; it far exceeded Air Force standards.

There is an upper limit to the rate at which a person can be spun before becoming dizzy or, in human factors terminology, losing orientation. Although people can function with spin rates as high as 10 rpm, such a spin rate causes them to suffer from a high degree of disorientation.

The second has to do with a Coriolis-like effect that may benefit certain space industrial processes but be a problem to humans who operate those processes. It's called *gravity gradient.*

Figure 13-7 illustrates this although it's not drawn to scale. The human occupant of this rotating spaceplex is held by centrifugal force to the inner

Satellite Rotational Rate Necessary to Produce Various G-Forces for Various Satellite Radii			
(Rotational rate expressed in rpms)			
Radius	G-force		
(feet)	1.0g	0.5g	0.25g	0.1g
5	24.17	17.09	12.08	7.64
10	17.09	12.08	8.54	5.40
15	13.95	9.87	6.98	4.41
20	12.08	8.54	6.04	3.82
25	10.81	7.64	5.40	3.42
30	9.87	6.98	4.93	3.12
40	8.54	6.04	4.27	2.70
50	7.64	5.40	3.82	2.42
60	6.98	4.93	3.49	2.21
80	6.04	4.27	3.02	1.91
100	5.40	3.82	2.70	1.71
120	4.93	3.49	2.47	1.56
140	4.57	3.23	2.28	1.44
160	4.27	3.02	2.14	1.35
180	4.03	2.85	2.01	1.27
200	3.82	2.70	1.91	1.21
300	3.12	2.21	1.56	0.99
400	2.70	1.91	1.35	0.85
1000	1.71	1.18	0.85	0.54

FIG. 13-8

surface of the outer wall of the spinning torus. The strength of the centrifugal force created by rotation depends upon the distance of the affected mass from the center of rotation. The farther from the axis, the greater the centrifugal force. The person's head is much closer to the axis of rotation and is therefore subjected to *less* centrifugal force. If the rotating part is 100 feet in diameter, the centrifugal force is equivalent to 1-g, and the human is 6 feet tall, the centrifugal force on the person's head is 6 percent less than at the feet. If the person lifts something, it becomes lighter as it's lifted.

This is called a gravity gradient although it's being created in this case by centrifugal force and is therefore pseudogravity.

A gravity gradient can also be created by Coriolis force. Suppose a person starts walking around the inside of the 100-foot-diameter rotating torus. If the person walks in the direction of rotation, he will be walking faster than the rotating part and therefore creating more centrifugal force, increasing "weight" by as much as 20 percent. If the person walks in the opposite direction against the direction of rotation, this is reversed.

To a human being, these variations in the magnitude and direction of the apparent gravity force can be unexpected and certainly quite unusual. They may therefore cause an individual to become quite ill and suffer visual hallucinations.

Centrifugal Force

Coriolis forces, gravity gradients, and problems with human tolerance to rotation rates have fewer physiological and psychological effects as the diameter of the spaceplex increases. This is why all centrifuged spaceplexes are large and why small facilities have to be of the weightless type.

The basic equation describing the magnitude of centrifugal force is:

$$f = mv^2r$$

where f = centrifugal force, m = mass, v = angular velocity in degrees per second, and r = the radius in feet of the rotating module.

The rotational rate in a small spaceplex can be tolerated if the required pseudogravity is reduced. Figure 13-8 shows the module rotational rate in rpm necessary to produce various g-force levels for various module diameters.

If people want to live in a 1-g environment, they should consider a very large spaceplex, which in turn means that many people will be living in it. It also may mean a longer wait to get in because the big rotating spaceplexes are more expensive, take longer to build, and therefore aren't as numerous as smaller ones.

If people want to live in 0.1 g's, the required spaceplex diameter would be 200 feet but the rotational rate would be 3.82 rpms, just at the limit where most people become uncomfortable because of Coriolis and gravity gradient effects.

Transition Problems

Another problem involved with living in a spaceplex that is partially centifuged—i.e., only part of the facility is rotating—is making the transition back and forth between weightlessness and pseudogravity.

The centrifuged portion must be entered through a complex compartment at the hub of the module. This is the place in a rotating module where the least difference exists between the relative speed of the rotating part and the non-rotating part. Depending upon the size of the rotating module, people then have to go toward the rim by climbing *backward* using a ladder against increasing pseudogravity force or using a personnel conveyor or elevator, if

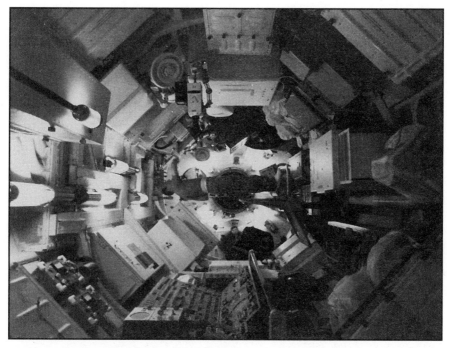

FIG. 13-9: *Inside a spaceplex module, all volume is used if practical because there's no such thing as "up." This was the interior of the NASA Skylab.* (NASA photo)

the module has them. Going from the rotating module to the non-rotating one is the reverse of this procedure, and it's quite a climb against pseudogravity all the way.

It may be difficult for people to adjust their equilibrium senses from the rotational world of pseudogravity to weightlessness and back. Some will suffer from motion sickness. Although most people are highly adaptable, some individuals may never be able to adapt; they may be able to live comfortably in a 0.1-g centrifuged module, for example, but cannot tolerate a daily transition to and from total weightlessness. No reliable test on Earth can determine this in advance.

Weightless Spaceplexes

The spaceplexes that exist continuously in weightlessness have different problems but are far more interesting and even more fun to live in if a person is highly motivated to go into space and live there.

Weightless spaceplexes can be designed and built much differently than those with centrifuged portions. They are far less complex because they don't have to withstand the structural stresses created by the large rotating modules. Nor are they affected by the gyroscopic forces of large rotating modules. This last creates problems with spaceplex attitude control if a specific orientation must be maintained with the Earth or the Sun.

Weightless spaceplexes can be smaller and more compact than rotating ones. Coriolis forces are no longer a factor because they don't exist. Furthermore, far more of the interior cubic volume can be used.

Living conditions in weightless spaceplexes will be as radically different from those in rotating ones as they are from Earth conditions.

However, until more data are accumulated on the true long-term (several years) physiological effects of weightlessness on people, a weightless spaceplex can't be considered to have a permanent crew. It must have a transient crew that is sent back to Earth every six months or so. In this regard, they resemble the nuclear-powered ballistic missile submarines of the US Navy with the Blue and Gold crews, each crew serving aboard for a few months, being replaced by the second crew and given R&D, and then returning to relieve the second crew. They are also analogous to offshore oil rigs where crews are also rotated on a regular basis.

The analogy to offshore oil rigs also holds true when it comes to supplying smaller weightless spaceplexes. Their life-support systems are probably of the open type requiring regular replenishment of oxygen, nitrogen, lithium hydride cannisters, and activated charcoal odor-control filters. Excess water created by human metabolism also will have to be removed because it may not be possible to vent it overboard into space without contaminating the environment in which a space industrial process must operate. Furthermore, someone must collect and dispose of garbage and other wastes by taking them to Earth or special space disposal facilities.

The biggest problem with weightless spaceplexes lies in the minds of their earthbound designers.

Very few people can think in three dimensions. As land animals, human beings evolved in a two-dimensional world and often experience difficulty when operating in a three-dimensional one. This is one reason why Earth has so many automobiles and so few airplanes. People find it natural to operate an automobile in its two-dimensional environment. Months of training and constant practice are required to successfully and safely operate an airplane in its three-dimensional world.

This "dimensional" shortcoming affects many of those who design weightless spaceplexes. Eventually, this shortcoming will disappear as more designers learn how to live in weightlessness by spending time there. After the first children are born in weightlessness and grow up to design future spaceplexes, the transition will be completed and these facilities will be truly designed for weightless living.

Designers who specialize in marine vessels or that almost extinct field of railway sleeping cars have come closest to knowing how to design for volume rather than floor space.

It's amusing to look at some of the mockups of the early space stations— Skylab, *Salyut*, and *Mir*—because the basic shortcomings in design for weightless living are apparent and reveal the earthbound thought patterns of the designers.

For example, the designers of the NASA space shuttle Orbiter derived its basic layout from the airliners of the time. The Orbiter had a "flight deck" and a "mid-deck." Both were laid out as if the Orbiter were an airliner flying horizontally through the Earth's gravity field.

None of the designers considered the basic problem of "crew ingress" (getting every crew member aboard and strapped into seats). When the crew gets aboard, the Orbiter is in a vertical orientation on the launchpad. Several quick fixes had to be introduced at the last moment to ensure that the flight crew could get to their seats that, in the launch orientation, were no longer on the floor but up on the walls.

Only one forty-inch-diameter hatch was provided on the left side of the mid-deck for the crew to enter and leave the Orbiter, even in an emergency. Another forty-inch-diameter hatch led from the mid-deck aft bulkhead into the Orbiter payload bay. Theoretically, nothing can be put into the Orbiter crew module if it exceeds this forty-inch size limitation.

Although the space shuttle's crew floats in weightlessness in orbit, they were still provided with ladders for interdeck movement as well as a stepping stool so that small people could "stand up" to the payload bay control panels located on the aft bulkhead of the flight deck.

Living in the Weightless Spaceplex

The best description of living in weightlessness is to say that it's similar to living under water without the water.

Living and working quarters in weightlessness are quite different from those encountered anywhere else because the quarters use *cubage* rather than just floor and wall space.

Even the most primitive weightless spaceplexes—including the ancient US Skylab—didn't use "dormitory" or "barracks" style living quarters. And, as spaceplexes got larger, individual compartments have been used in increasing numbers. Among the reasons individual living quarters or "rooms" are used include both safety and structural strength.

Dividing a large internal volume into smaller volumes with bulkheads permits structural loads and forces to be routed through multiple stress paths in the same manner as a jet airliner is designed with multiple load paths. Although engineers have to make design trade-offs between single-wall self-supporting or monocoque structures and multicellular enclosures, the internal bulkheads of smaller personal compartments help distribute the loads and, with only a small mass penalty, also provide pressure integrity and safety similar to the watertight compartments of an ocean-going vessel. If pressure is lost in a single compartment because of, say, penetration by a rare meteor large enough to broach the external meteor shield, the affected personal compartment can be sealed off quickly to prevent the entire internal atmosphere of the spaceplex from leaking into space.

Another reason for building spaceplexes with personal living quarters relates to human needs for comfort and privacy. These factors lie behind the way humans build their dwellings on Earth. Although primitive and poor cultures still have dwellings in which a single room is used by many people, they rapidly build new multi-room dwellings or add onto existing ones as quickly as their means permit it. The ancient Sumerians built houses with separate rooms as long ago as 5,000 B.C.

Therefore, most of the spaceplexes have private or semi-private living quarters in which people may "double-up" to share with someone else.

Although the way that personal living quarters in space is designed may at first seem strange, people quickly grow used to it. At first, their earthbound ways of thinking may tell them it's too small. However, the walls, the ceiling, the floor, and every nook and cranny can be used in the weightless environment.

Furniture as it exists on Earth isn't needed in weightless living quarters. Furniture is primarily designed to *support* a human being against the force of gravity. Zero-g furniture exists, but it serves different functions than earthly types.

There is no bed in private quarters. People have a choice of sleeping in a sleep sack or under the gentle restraint of belts, either of which will keep them from floating around the compartment in their sleep and possibly banging into something that could cause injury.

Chairs don't exist because those earthly artifacts are useful only in a gravity environment.

The working desk that folds down from the bulkhead may have knee or leg

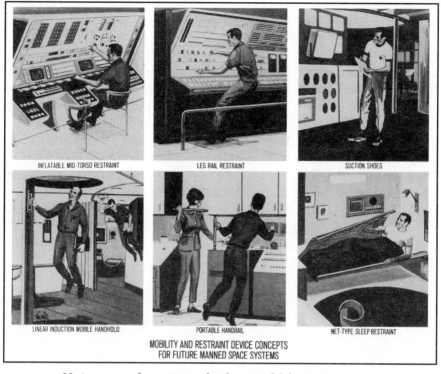

INFLATABLE MID-TORSO RESTRAINT   LEG RAIL RESTRAINT   SUCTION SHOES

LINEAR INDUCTION MOBILE HANDHOLD   PORTABLE HANDRAIL   NET-TYPE SLEEP RESTRAINT

MOBILITY AND RESTRAINT DEVICE CONCEPTS
FOR FUTURE MANNED SPACE SYSTEMS

FIG. 13-10: *Various types of restraints and aids are useful for moving or staying in one place in a weightless spaceplex.* (NASA photo)

restraints under it so a person can tuck up against it for working. Although this desk is similar to the eating tables in the common room or wardroom, people normally don't eat in their personal compartments. The personal desk may be used for writing, although that primitive form of communication and data keeping is obsolete because each room has a computer terminal. This gives access to the spaceplex general purpose computer with network links to data banks in the facility and on Earth. Paper isn't made in space and actually becomes an item of great value.

The compartment has its own entertainment center that may be incorporated into the computer terminal.

It has storage lockers for personal items such as clothing, personal hygiene items, and even ultra-valuable books.

Some compartments have their own bath and toilet facilities and some a separate set of facilities that is shared between compartments. The size, complexity, and age of the spaceplex determine how personal hygiene and public health facilities are arranged and located. The newer and larger the spaceplex, the more luxurious the bathrooms will be. Even in the smaller and simpler spaceplexes, these may seem to be primitive but are far better than people had on any terrestrial frontier in the nineteenth century.

If the personal compartment doesn't have a viewport, the common room certainly does. People like to look out the window in space and watch the world go by. So all space complexes have as many of these as possible, in spite of the fact that it's impossible to keep them from leaking a little bit. View ports are part of the human side of space living because of human needs.

In many respects, living in space isn't that much different from living on Earth except for differences in the environment. These differences make space living unique, of course. And in the same vein, living with other people in space is at once similar and different from Earth.

FIG. 14-1: *Working in a space factory.* (Art by Sternbach)

CHAPTER FOURTEEN

SOCIAL ASPECTS OF SPACE LIVING

Before people began to live and work in space, early space advocates and dreamers spent a lot of time and effort trying to figure out the social aspects of the spaceplexes (they called them "space colonies") they envisioned. Most of these people were physical scientists and technicians with little if any knowledge, let alone personal skills, in social areas beyond their specific professional fields of expertise. Their work on space colonies was technically thorough. However, it was far in advance of the willingness of bankers, financiers, venture capitalists, and even Congress to spend the money to create these space colonies. The space colony advocates had neglected (or perhaps didn't know how) to answer such important questions as:

"Where's the money going to come from?"
"How much is this going to cost?"
"What are all those people going to *do* in space colonies?"
"What's the return on investment?"
"What's in it for me?"
"Why do it at all?"

The Space Utopias

Economic issues aside—because none of the advocates knew how to answer such economic and financial questions—other questions had to do with the social aspects of the space colonies. The visions of the space colony advocates tended to be strongly utopian and collectivist. There is nothing wrong with having a utopian dream because it gives people something to work toward in terms of long-term goals. The collectivist portion, however, left people with the unanswered questions, "Who does what for whom and who pays for it?"

A lot of study, speculation, and outright wishful thinking was thus done by space advocates who attempted to lay out a "perfect" space colony or spaceplex. They did an excellent job of selecting the size, shape, internal layout, materials, structural design, and life-support systems. But they also treated the potential inhabitants as human machines in perfect accord with the tenets of materialism.

The screening and selection of people to dwell in space by these utopians was seriously considered—often in near-complete ignorance of human nature and the history of analogous settlements on Earth. It was suggested that people be chosen on the basis of intelligence (the intellectual or "Mensa" utopia), physical fitness (the athletic utopia), psychological fitness (the introverted utopia), compatibility (can't we all get along?), reliability (you have my word on that), and a host of other parameters, most determined on the basis of somewhat flexible standards and testing results that can be interpreted and therefore biased in any direction the selection committee desires.

The justification for such a selection process was the belief that the cost of getting people into space and keeping them alive there would be high. It was felt necessary to screen out the deadwood. However, the cost justification has vanished with the demise of the government monopoly space program and the ascendancy of economical commercial space activities. And if anyone really knows how to make this kind of selection, they can make far more money here on Earth because there are thousands of corporations that would be eager to embrace this knowledge and pay well for it.

Americans seem to forget their own history, to say nothing of the history of other parts of the world. Many of their progenitors came to America because they *had* to. They were debtors running from bill collectors and prison terms, criminals on the run from police for various offenses including political and religious heresy, and general misfits in the European cultures. The entire continent of Australia was a penal colony to which English people could be banished with little or no chance they'd ever show up in England again to make more trouble.

Every generation develops its own concept of utopia. In the last few centuries with people becoming richer and more in control of the forces of nature, many groups of people managed to achieve their utopias. In so doing, however, they discovered that reality turned out to be a far cry from what they'd envisioned when they'd started. They learned that achieving a utopia—or the creation of any sort of social institution—requires an unprecedented and un-

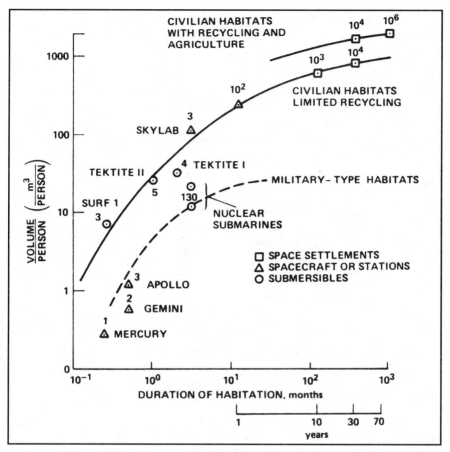

FIG. 14-2: *Chart showing the personal space and volume required by people.* (NASA)

anticipated amount of hard physical and mental work accompanied by worry, anxiety, panic, controversy, arguments, tragedy, accidents, and far more deaths than any of them would have found acceptable at the start.

This basic description applies to all human frontiers—past, present, and future.

People don't do their homework in such matters. In a way, this is probably good. If they really knew at the beginning what problems, troubles, set-backs, and conflicts they were going to have, they might not have embarked on the scheme in the first place.

Without trying to over-study the situation in advance, prudence would dictate that those who want to forge the new space frontier should carefully study the lessons learned by their forebears on the old frontier. Even when people try to do this, they don't succeed. Each person and group learns what it considers applicable, neglects the rest, and forges ahead with a biased and distorted view of frontiersmanship based on legends, folk tales, fiction, motion pictures, television shows, and wishful thinking. This leads to mistakes that increase the monetary and human costs of opening a new frontier.

Space is no different, even though space offers an opportunity to start afresh without the ghosts of past mistakes to haunt the pioneers. But they should know what the ghosts of the past are. Most of them are social in nature.

Institutions

At this point, the word "institution" needs to be defined because many readers use the word in the context of their own experience and education rather than the context in which it's used here.

- An institution is a group of people.
- It's organized for a purpose.
- The people in it follow rules agreed upon among themselves.
- It has a structure that, in its simplest form, has a leader and a few followers.
- It has internal protocols so that people know who's in charge and the relative ranking of individuals within the group's organizational structure. These protocols include rewards for activities that benefit the group as well as punishments for breaking the rules.

In line with this, the group may develop a means of keeping score so that those who are rewarded may be recognized in a manner commensurate with their beneficial activities and those who must be punished may be dealt with in a just manner.

It has procedures for admitting new people to the institution and for members to leave, voluntarily or forcibly.

These elements involve the institution's internal politics; it also will have rules that guide its relations with other institutions.

It's important to remember that an institution is a group of *people* and not a collection of ideas, rules, practices, or customs.

Space Justice

The method of enforcement of the spaceplex or spaceship rules also depends upon how these facilities are operated and governed. Small spaceplexes don't have dedicated police forces.

Space law began in the 1950 decade almost as an academic exercise among lawyers who were interested in space. Originally, before people began to live and work extensively in space, it became involved with such impersonal problems as communications frequency assignments in space, orbital slot allotments for communications satellites, liability for damage from spacecraft that crashed on another nation, and, to a limited extent, the extension of the law of the sea and the law of the air into space. The biggest problem with early space law was that lawyers tried to anticipate technology. They established space law principles based upon what amounts to technological speculation, much of which turned out to be wrong, incomplete, and inapplicable to what really happened in the commercial space age.

True space law began when people started to live and work in space because the basis of all legal codes and actions is the nonviolent resolution of conflicts between people.

The legal systems that exist in various spaceplexes are straightforward extensions of whatever legal systems existed in the institutions of the earthbound companies or the nation/states that established the spaceplexes in the first place. These legal systems are slowly evolving and changing to meet the needs of the various space cultures.

Culture

An institution also is not a "culture," a word with about as many definitions as there are social scientists. Culture is the *way people live* in or with an institution. Culture is transmitted from generation to generation by learning. It includes the relationships between individuals and the institution, the way they work within the institution in handling matter and energy, and the way they operate in handling symbols such as speech, music, the visual arts, and others relating to communications. Culture is thus the total sum of what people do as a result of being taught these things.

The Primitive Social Sciences

Early space colony advocates attempted to tackle the many social aspects of space living using the meager and inadequate tools of the "social sciences" of the time.

These adjectives are deliberately used here because the social sciences have yet to reach the point where they are predictive sciences such as physics, chemistry, and astronomy. Progress is being made, but much of it is coming from beyond the walls of the social sciences redoubt. Because large, complex social systems are chaotic, they may never become predictable.

However, investigators are bridging the gap between the physical, biological, and social sciences in an unsuspected area: control mechanisms for complex systems. This has led to the development of a field of scientific/mathematical inquiry called "semiotic analysis."

Semiotics is defined as that area of knowledge expressible in some form of language. It has two sides: (1) the theoretical foundations, and (2) the practical applications. At the time of this writing, early in the evolution of this new field of scientific and technical investigation, theory and practice have little to do with one another. An enormous difference exists between theoretical semiotic theory and practice of natural languages and practical semiotic theory and practice of formal languages. These latter include logic, mathematics, and computer programming languages.

On the practical engineering side, computers are simulating more and more semiotic activities of the human brain. In some cases, the technological systems are so complex that they couldn't function without computer controls. In practice, most engineers worry about software reliability and don't care about the philosophical or physical functions of semiotic control theory.

On the academic or theoretical side, the theory of semiotic controls and how they're related to natural laws is a fundamental cause of controversy. For example, in physics the problem is how to decide when a measurement is completed. In biology, the issue revolves around how much of the structure and coordination in organisms is the result of semiotic control of the genes and how much is the result of self-organization. All these theoretical issues are special cases of a basic problem: What is the difference between information and knowledge, and how do we know the difference?

An organic cell, the human body, or a collection of human bodies in a social institution exhibit the same characteristics. They all operate as open systems in which inputs from individual elements, often made by what appear to be random processes, can affect the overall performance of the system and even start it in a new direction. A major area of interest is in control systems for such complex systems. Social institutions are control mechanisms for complex multi-human systems. Many social scientists and nearly all politicians haven't awakened to these new concepts yet.

New tools have recently appeared. One of these is chaos theory, something people intrinsically understand in spite of its novelty. "Of course! That's nothing really new! Why didn't I see it before?"

All of this plus the accumulated knowledge of history will help people develop and evolve some unique social institutions in space. In turn, the social sciences will benefit and be able to make the transition to a predictive science.

But that point hasn't been reached yet. It's extremely difficult to get most social scientists to admit this. However, social scientists cannot yet predict the future behavior of the United States of America. If they could, there would be no "boom-and-bust" cycles and the nation could be truly "managed" (which may or may not be a good thing, depending upon one's personal ideology).

A Successful Social Institution

One group of people actually designed a large social institution that turned out to be capable of considerable growth and has thus far lasted successfully for more than two centuries. This was a collection of fifty-five men who were not social scientists but military officers, financiers, public officials, merchants, traders, lawyers, and planters who met from May to September 1787 in Phila-delphia to thrash out a Constitution for the United States of America. For the next two centuries, the people of the United States—who weren't social scien-tists either—spent a lot of time and effort making that Constitution work. Even after two hundred years, it isn't a perfect document telling people how to operate a perfect form of government. It requires a high level of education and a lot of work to use it.

As historian Dr. Daniel Boorstin pointed out in his book, *The Republic of Technology*, it's an experimental document that establishes an experimental institution that always changes but is based upon centuries of know-how that mostly told the members of the Constitutional Convention what *not* to put in the document. Few people know that many of the principles behind the Con-stitution, such as a bi-cameral legislature and the three arms of the govern-ment, to say nothing of much of the coded law behind it, came from ancient Sumer more than 5,000 years ago. Thus, it followed the main line of develop-ment in human history.

Far too many of the early "social engineering" proposals for space colonies took the easy way out when specifying as the primary government social insti-tution a centralized, authoritarian, bureaucratic, and totalitarian political and governmental scheme. This may be the simplest approach but, as has been seen in the latter years of the twentieth century, it's not the most durable or workable.

Generally without the sort of background and knowledge of history and philosophy that the signers of the Constitution possessed—they were learned men for their time—the space colony advocates "designed" populations for their proposed space colonies and worked out criteria for the selection of people who would be allowed to live there. They also designed forms of gov-ernment, codes of law, economic systems, and even monetary systems, most of which had been historically tried and failed, which they'd have known if they'd read their history books.

Others went to the opposite extreme and designed anarchistic "let it all hang out" romantic communal type institutions based upon the twentieth century's romantic movement that began in the 1960s. (It had been preceded by similar romantic movements in the previous two centuries.)

Both institutions may work under specialized conditions and for short periods, but not for the reasons understood by the space colony advocates.

Institutions and Culture in Space

Social institutions in spaceplexes will evolve beyond those of Earth after which they were patterned. But they will start, as did the United States of America, with a basic foundation of past and existing institutions that people know will *work*.

Space cultures are still evolving. It takes a generation or more for a new culture to emerge. In the meantime, people will live with the earthly cultures they've taken with them into space, adapting them to work in the new circumstances. People who have been in space will teach newcomers and children what they've learned to be workable and unworkable.

Social institutions will exist in space because people *must* organize themselves in a new environment that is somewhat hostile and deadly to the rugged individualist. Actually, this is also true of any place on Earth. The places where the rugged individualist can exist in total isolation, much less live well, have been disappearing for centuries because, even on Earth, the cooperation of other people in other institutions is necessary to support a life style other than the most primitive. Many of the followers of the romantic movement of the 1960s discovered this when they eschewed technology and decided to go back to the simple life and live off the land. It turned out to be a very primitive existence. Many of them didn't go that far or came back to continue to espouse their fear and dislike of technology and its institutions by using these to communicate their ideology and recruit followers.

Although the life style in space will be adequate and comfortable, and although space civilization, cultures, and institutions will continue to evolve, few if any spaceplexes will operate with any resemblance of a utopian civilization. Utopias don't exist on Earth or in space. They are ideals. Most spaceplexes are sponsored, paid for, built, and operated by institutions on Earth. They may be public, quasi-public, or "private." But their institutions and internal social structures will be pragmatic copies of similar ones on Earth.

Paramilitary Spaceplex Organizations

Early, small space outposts are likely to be quasi-military, authoritarian hierarchies that have derived from the paramilitary nature of the early government space programs of the United States and the Soviet Union. Although overtly "civilian" in nature, these were organized, staffed, and managed by people who were trained in the military services or who had spent most of their lives working for the armed services. These tax-supported space programs were justified on the basis of national prestige and therefore national security. Early space launch vehicles were modified military missiles. A large percentage of the astronauts and cosmonauts had been military aircraft pilots. (In fact, in most of the world today, the fact that a person is an airplane pilot automatically means that he or she was trained to fly by the armed services of their country. "Once a soldier, always a soldier." The author, a private pilot with no military service, encountered this in eastern European countries.)

A quasi-military organization isn't necessarily a bad way to run a spaceplex. Naval forces of the world over many centuries developed highly successful maritime institutions relating to the adequate performance, safety, health, and comfort of large groups of highly trained people living and working in isolated, dangerous environments. They are pragmatic experts in the psychology of people living closely together in isolated, dangerous places for long periods of time.

Technobureaucratic Spaceplex Organizations

Some spaceplexes may run on a system of technocratic centralized control somewhat similar to the quasi-military spaceplexes but far more highly concentrated in the technical area because of the purposes of the spaceplex. This is especially the case where the spaceplex happens to be a construction base or a research-and-development laboratory. Again, this institutional form of organization isn't necessarily bad or restrictive.

If the spaceplex is under the control of a large domestic or multinational corporation or a consortium of companies, the organization of the spaceplex may mirror the management techniques of the parent organization. One of these forms may be a participative meritocracy.

Participative Democratic Spaceplex Organizations

A few spaceplexes may have a social organization and government that is the self-organized popular participative democracy operating by town meetings. Such an organizational form requires either elected voluntary leaders who must govern in their spare time away from their everyday work in the spaceplex or the employment of specialized hired managers. It is the difference between the government of a small New England village and that of a large city.

The Critical Factor

People have developed many institutions and forms of government that work. Some work better than others. Some work well only in special circumstances. However, history reveals one that doesn't work very well or for very long. This is the institution in which the individual member isn't responsible to the group or the group's leaders, or where the leaders aren't responsible to the group or to some higher and accessible appellate authority for any action or activity of either a committed or omitted nature. This is absolutism.

Absolute power does not corrupt absolutely. Absolute immunity does.

Emerging Space Institutions

An institution cannot remain static and survive. An institution developed for one set of circumstances cannot be expected to work successfully in a widely different set of circumstances. This is especially true of the emerging social institutions of space. While institutions may start by taking one of the forms discussed or others, they will evolve as people develop their own sense of community and learn how to organize better for their personal and collective survival in a new environment.

People who originate in the modern superindustrial cultures of the United States and Europe may have to make some difficult personal decisions when deciding to go into space to live and work. Most of these involve assigning priorities to personal freedoms. The most important element of our evolving social institutions on Earth is the preservation of personal freedom of choice consistent with and contributing to the survival of the institution. This may be and has been looked upon as a sort of implied acceptance of a social contract. Probably the most important freedom that should be preserved in space is that which will allow a person to leave one spaceplex to seek a life style in another that promises to offer more of what is desired.

Social Revolution in Space

Some scenarios forecast revolutions of the sort in which the thirteen colonies of North America broke away from the English crown. Deemed less likely is the sort of revolution that's taken place far more often on Earth in which the *ancien régime* is replaced by a new ruling group. The sense of cultural integration in a spaceplex is far more likely to lead to a separatist type of revolution than a move to "throw the rascals out."

However, revolutions that sever ties with earthbound institutions aren't foreseen until far more people are living in far larger spaceplexes with closed ecologies. It's not a good idea to revolt and break earthly ties if a spaceplex depends upon oxygen, nitrogen, food, and other supplies being ferried up from Earth on a regular basis. Those are valuable commodities, but they aren't the only ones that take on new values in space.

Valuable Items in Space

Regardless of the type of social institution in a spaceplex, commonplace earthly items will turn out to have greater value in space if they've been brought from Earth. Such items are rare and expensive in some spaceplexes. On the other hand, many expensive items on Earth may be relatively commonplace in space and therefore cost less.

Paper is an example. On Earth, paper must be made from wood pulp. Gravity and a lot of water are required in the papermaking process. Therefore, since it can't be made in space, paper of any sort must be transported to a spaceplex from Earth. Paper is heavy, It has a density averaging between 44 and 72 pounds per cubic foot (water has a density of 62 pounds per cubic foot).

A box of facial tissues or a roll of toilet paper therefore has a higher value in space. People who live in space quickly learn to use all sorts of paper sparingly.

As a result, paperwork is rare in space. Paper books don't exist while books on CDs and stored in other electronic media are commonplace and cheap. Letters are rarely used, being replaced by e-mail sent through computer networks. Communications are very cheap in space.

Clothing is also valuable because the technologies of spinning, carding, weaving, and dyeing are difficult to adapt to the weightless space industrial environment. These are old technologies on Earth and they require gravity. They may be among the last adapted to the space environment.

In the meantime, people must have clothing to wear if for no other reason than to prevent chafing. It's imported from Earth and is expensive. Extensive space wardrobes are rare because of this expense. Fireproof or fire-resistant clothing may be required in some compartments or sectors of a spaceplex. Other types of cloth won't be permitted in certain areas because of flammability, high-temperature characteristics, or outgassing qualities. Shoes become protection against people stubbing their toes. Necklaces and neck chains won't lie on a person's chest in weightlessness and may become tangled with some part of the facility, becoming a noose.

Money in Space

Value leads to the concept of money that is yet another interesting new social aspect of living in space. People are well-paid for working in space because they do jobs where the potential hazards are great. But where can they spend the money they earn?

In some spaceplexes, money doesn't exist because there's nothing to buy and no place to spend money. Small spaceplexes that are run in a paternalistic manner fulfil all the needs of a person living there. Salaries are banked on Earth.

In other spaceplexes, money exists only in computers that keep track of who owes how much to whom, a growing trend even on Earth. These spaceplexes operate like the corporate towns in the mining districts of Earth or petroleum workers' towns in the Middle East. In these "company habitats," a person's account is credited regularly with a "salary" and then debited when food or other items are "bought" in the cafeteria. It also may be debited for the use of air, water, electrical power, waste-handling, and use of a personal compartment. Whether this is done at all and the extent to which it's carried out depends mostly on the way the organization's bookkeeping is set up and how comptrollers, accountants, tax collectors, and tax lawyers interpret the then-current tax codes. In such "moneyless spaceplexes," barter and in-kind trading is the most common form of value exchange.

Other spaceplexes issue pseudomoney in the form of chits or tickets for meals, water use, baths, air-lock cycles, and other activities that must be kept under control because of costs, logistics, or the accounting system used by the institution. This chit system is widely used on Earth in a variety of organizations ranging from schools to international sporting competitions. A ticket or chit is pseudomoney because, unlike money, once it's used it becomes worthless thereafter. However, if it hasn't been used or is only partially used, it has value. Tickets and chits therefore are items of value to trade, much like poker chips in a poker game.

Other large spaceplexes with many people living in them have a definite form of money. This may take the form of some sort of readily identifiable and difficult to duplicate token. Paper may not be used at all because it's too valuable.

The Great Social Experiment of Space

The development and evolution of space institutions and cultures is an ongoing and growing process that will provide social scientists with the best hard data they've ever had. It will add rigor to their body of knowledge.

Watching this great space social experiment unfold will be as fascinating as what people in space will learn about the Universe itself.

As continued development of the space frontier leads to more and more settlements and more and larger spaceplexes, additional institutions will be formed just as they have been on Earth. Everyone on Earth belongs to several institutions—church, service club, social club, hobby club, professional society, tribe, local and state governments, and even international organizations. People who go into space to live and work will take these memberships with them, either adapting them to the evolving cultures of space or forming new ones using the Earthly ones as models. There will be a rising degree of community identity and integration that will inevitably result in a growing desire on the part of people to separate their institution's decision-making powers and governmental organizations from those on Earth. There may eventually be some manner of "interspace" organization or organizations equivalent to the interstate and international groups on Earth.

Social Experiments in Space

Other social institutions are expected to develop, evolve, and mature in space as time goes by and more people live and work there.

Space offers the social scientist one of the best opportunities in history to engage in long-term studies of human interactions, how institutions behave, and how both individuals and institutions react and interact to changes in their environments and the situations in which they exist. Such studies are difficult on Earth because of the age of terrestrial institutions and the ongoing interactions between people and institutions. No "social experiment" can really be isolated or even semi-isolated so the investigators can get a solid grip on all the variables.

Although this is still the case to some extent in space because of the proximity of spaceplexes to Earth, the quick and easy communications with Earth, and the dependence upon earthly sources of supplies, the evolution of future independent, self-sufficient space institutions in locations remote from Earth provides the social scientists further opportunities to bring rigor to their field by true social experimentation.

Space is not only the final frontier for technology but for the social sciences.

FIG. 15-1: *Spaceplexes will become quite large and be worthy of the title, "space complexes."* (Art by Sternbach)

CHAPTER FIFTEEN

THE GIANT LEAP

W<small>HEN NEIL A. ARMSTRONG</small> stepped onto the surface of the Moon at 02:56:20Z on July 21, 1969, his now-famous words echoed around Planet Earth. He knew what he was saying, and he knew the deeper meaning of it. It took more than a decade for people to understand what Armstrong meant because he wasn't speaking for the history books as much as for the generations to come who would live and work in space, extending humanity's ecological niche and expanding it to the stars.

By going into space to live and work, people are taking part in something as vitally important to the future of the human race as anything that has happened in history.

People have been building civilization for 5,000 years or more. Civilization and its institutions are still being built. People take this heritage with them when they go into space. But the civilizations built in space may be as different from what has been developed on Earth as those first primitive cities of Sumer were from the tribal villages that preceded them.

Civilization is, literally, the art of living together in cities. That's what the root words mean. The concept of "city" has changed and expanded over the centuries. Historians Ariel and Will Durant defined civilization as "social order promoting cultural creation. Four elements constitute it: economic provision, political organization, moral traditions, and the pursuit of knowledge and the arts. It begins where chaos and insecurity end. For when fear is overcome, curiosity and constructiveness are free, and man passes by natural impulse toward the understanding and embellishment of life."

What people are doing in space is exactly that—providing for themselves and others in the economic area by their work and production of items of value to others, organizing and operating the necessary institutions for political and cultural growth, and pursuing knowledge and the arts.

A space civilization will not spring up immediately as the Greek goddess Athena sprang from the forehead of Zeus. It must be created. This takes time. Civilization on Earth has taken at least fifty centuries to reach its present globe-spanning level. But it will be possible to create a space civilization faster because of the great storehouse of historic knowledge that provides a foundation plus the ability to communicate it at the speed of light.

The evidence in this storehouse of knowledge points to the fact that the human race is in an era of transition, the greatest age of change that has taken place in the last fifty centuries. The late Dr. Herman Kahn observed, "Two hundred years ago the human race almost everywhere was few, poor, and largely at the mercy of the forces of nature whereas two hundred years hence, barring some perverse combination of bad luck and bad management, the human race should be almost everywhere numerous, rich, and largely in control of the forces of nature." The development of space civilization is part of this transition. And because of Kahn's observation, another aspect of the civilizing process should be added to the list: learning how to be rich because it means more than just having a lot of money.

To get some perspective on this, consider: It's been about two hundred years since people began to fly in the air, albeit in balloons. Before that, people left the surface of the Earth only when they jumped from a higher hill to a lower one. No one could ascend into the air like a bird. The first flight by François Pilâtre de Rozier in the hot-air balloon built by the brothers Jacques Etienne and Joseph Michel Montgolfier took place near Paris, France, on October 15, 1783. Benjamin Franklin, who was present as the United States' Ambassador to France as well as a scientist in his own right, is reported to have said to a querulous spectator of the event, "You ask of what use is this? Sir, of what possible use is a newborn babe?"

Franklin, one of the most intelligent and educated men of his time, couldn't possibly have imagined what takes place every day two hundred years later at John F. Kennedy International Airport where thousands of people embark on short and comfortable journeys across the Atlantic Ocean—a journey that takes about six hours in contrast to the six weeks of hazardous sailing ship passage Franklin endured to get to France.

Many "right-thinking" people consider any other person's desire to go into space as an attempt to escape the "harsh realities" of living on Earth. But the "harsh realities" aren't on Earth; they are in space where no escape from them is possible without knowledge and technology. Living in space is not easy and won't be for decades to come.

Space is an environment that is just as dangerous, hazardous, and deadly for humans as most of the natural environments on Earth—the broad savannahs of central Africa, the rain forests of the tropics, the sprawling dry deserts, the cold mountain plateaus, the seemingly endless expanses of the oceans, the lifeless tundra of the polar regions, and the envelope of air that surrounds the planet. Space is as different to live in as dry land was to those proto-amphibians who somehow got stranded on the beach when the Silurian tide went out—and survived.

A magazine cartoon once showed a fish crawling up the beach on stubby fins and saying to other fishes still in the water, "Why? Because this is where the action is, baby!"

Even though some people may go into space for only a few months at a time, others may live there for extended periods. The numbers of long-timer space people will increase as the knowledge of space medicine and space living grows and matures through experience.

As this book has shown, a human being is well-equipped for this task, with a physical body and a large reasoning brain that has evolved and overcome all the hazards of the terrestrial environments. Humans are, as a result, members of the most powerful species on Earth. People in space will use these faculties to do the same in space if they know and understand what they are and where they came from.

People are also well-equipped from the social point of view. Their ancestors learned to live together in the teeming cities of Europe, Asia, and America. They will learn how to live together in the far cleaner and more thoughtfully designed cities of space, the spaceplexes. These will look nothing like the cities of Earth and probably nothing like forecasters envisioned them. New institutions will come into existence. Eventually, self-sufficient cultures will gradually emerge in space. They will no longer be tied to Earth for consumable supplies but will be self-supporting, even producing surpluses that will be traded elsewhere in exchange for other items of value.

People not only take four million years of human development with them but also the accumulated knowledge of the state of the art in every area of human endeavor.

Most of all, people who go into space to live and work are making the first moves in a long and important task: getting humanity's eggs out of one fragile planetary basket.

If, as a result of Kahn's "perverse combination of bad luck and bad management," a catastrophe of an astronomical (asteroid impact) or human (thermonuclear war or plague) nature occurs, human genetic and intellectual seeds may survive in space in people ready to return like gods from the skies to rebuild a shattered world.

The next step will be the expansion of humanity to the point where it could survive even if the Sun itself explodes or dies. But such an expansion of the human race beyond the solar system involves technology as yet unavailable and still highly speculative. People will have plenty to do in the next few hundred years right in their own "back yard," the solar system.

Although people go into space for personal reasons, these impersonal ones and the philosophy behind them are important to the human race as it grows into maturity as a species.

ACHNOWLEDGMENTS

WHEN ONE HAS BEEN WORKING for and thinking about living and working in space since 1950, a *lot* of information is garnered from books, magazines, professional meetings, space advocate conventions, and personal conversations with uncountable numbers of people. I've pruned my library at least four times because of the requirements to relocate. I've written two previous books on this subject plus at least six contemporary "hard" science-fiction novels involving people doing things in space in the twentieth and twenty-first centuries. I'm not trying to take credit for the work of others by not attributing it; it's only because *there's so much of it!* Thus, it's totally impossible to mention every book or technical paper I've read, all the people I've talked with, and all the inputs I've received that have allowed me to write this book. This is why the book has no bibliography; it would probably be longer than the book itself! I can single out some people who have had a major impact upon my work and my background, starting with my father who was an M.D. and going on to Robert A. Heinlein, Arthur C. Clarke, Gene Roddenberry, Dr. Willy Ley, Dr. Herman Kahn, Frederick C. Durant III, Dr. Clyde W. Tombaugh, Daniel O. Graham, Krafft Ehricke, Dr. Gerard K. O'Neill, Dr. John Paul Stapp, Jacqueline Cochran Odlum, Dr. William O. Davis, Lewis J. Stecher, John W. Campbell, Jr., and Dr. Wernher von Braun. I know I've neglected to mention others, and I apologize for inadvertantly not naming those people who, like those on the list above, I was and am privileged to have as my friends as well as colleagues and fellow space enthusiasts.

If the twentieth century is remembered for any good and wonderous reasons, among them will certainly be that this was the century when people awakened to the fact that they lived on a small, finite planet and could—if they wanted to do it badly enough for money, survival, or many other reasons—expand throughout the solar system to live there.

INDEX